ION NITRIDING AND ION CARBURIZING

Proceedings of
ASM's 2nd International Conference on
Ion Nitriding/Carburizing
Cincinnati, Ohio, USA
18–20 September 1989

Edited by
T. Spalvins and W. L. Kovacs

Sponsored by
ASM INTERNATIONAL®

Co-sponsored by
NASA Lewis Research Center

Published by

ASM INTERNATIONAL®
Materials Park, Ohio 44073

Library of Congress Catalog Card Number: 89-081481
ISBN: 0-87170-362-9
SAN: 204-7586

ASM INTERNATIONAL®
Materials Park, OH 44073, USA
(216) 338-5151

Printed in the United States of America

Conference Co-Chairmen

T. Spalvins
NASA Lewis Research Center
Cleveland, Ohio, USA

W. L. Kovacs
Elatec, Incorporated
Andover, Massachusetts, USA

International Organizing Committee

O. T. Inal
New Mexico Institute of Mining
  & Technology
Socorro, New Mexico, USA

S. C. Kwon
Korea Institute of Machinery
  & Metals
Changwon, South Korea

W. Rembges
Klockner Ionen GmbH
Leverkusen, FRG

M. Gantois
Laboratoire de Genie Metallurgique
Nancy-Cedex, France

F. Hombeck
Klockner Ionen GmbH
Leverkusen, FRG

K. T. Rie
Technical University of Brunschweig
Brunschweig, FRG

R. Grün
Plasma Technik Grun GmbH
Siegen, FRG

P. Lidster
Exactatherm, Ltd.
Scarborough, Canada

G. Kiss
Cleveland Twist Drill Co.
Cleveland, Ohio, USA

FOREWORD

Conference participants seem to characterize the flavor of a conference and have a sense of lasting remembrance on one or two key incidents.

In the First Conference on Ion Nitriding held in Cleveland in 1986, conference participants remember the unexpected charm and beauty of Cleveland in warm Indian Summer weather, the unexpected delight of a beautiful trip down the Cuyahoga River, and a sense that ion nitriding "had arrived" and the period of "art" had ended. There was a sense of expectation and a desire to make the Ion Nitriding Conference a regular event.

At the Second International Conference on Ion Nitriding/Carburizing, 40 papers were presented, and the program was divided into 6 sessions:
* Glow Discharge Diagnostics and Comparison of Related Ion Processes to Ion Nitriding
* Characterization/Analysis of Ion Nitrided/Carburized Surfaces
* Mechanical and Tribological Properties
* Industrial Applications and Performance
* Processing and Parameters, Design and Production Equipment
* Ion Carburizing

Participants at the Second Annual Conference will agree that one of the highlights of this Conference was the group interaction that developed at the open panel discussion held on Monday evening. (It is indeed a rare event when conference participants will reconvene after a heavy day of technical papers for a two hour panel discussion with virtually full attendance through the entire discussion.) Once again, the participants felt that the ion nitriding/ion carburizing processes have begun to achieve recognition emerging as one of the leading case hardening technologies for the next decade.

Conference participants could be divided along the following lines: scientists (school and research institutions), captive users for production parts, heat treat shops offering plasma services on a toll basis, equipment manufacturers and prospective users. In the Second Annual Conference , it seemed that a true balance was achieved among all of these groups.

In the field of academic research, the mechanism of plasma treatment as differentiated from traditional diffused case hardening processes is being better understood. Added to this is the emerging field of "multi-process" treatments. (These are treatments whereby plasma diffused processes, such as ion nitriding and ion carburizing are used in conjunction with other processes such as ion implantation, and coating processes (chemical vapor deposition and physical vapor deposition). Unusually high wear resistance and other enhanced physical properties are being reported for these multi step processes. These results created a true sense of "excitement" and interest among conference participants. In the future, these multi step processes will emerge as production processes at a much more rapid rate.

Equipment manufacturers seemly have settled their differences and grudgingly acknowledge and recognize that virtually all ion nitriding equipment offered for sale is of high quality and that individual user preferences dictate specific options and features desired. Minor controversies as to the benefits of using auxiliary AC heating to enhance temperature uniformity and reduce sputter time seem to be accepted, as well as the fact that both conventional SCR fired DC power supplies, rectified AC to DC power supplies and pulsed high frequency power supplies each have advantages and disadvantages. The user may select a power supply and controls based upon his specific applications (material, treatment, geometry, pre-treatment, etc.). Equipment manufacturers are now turning their attention to supplying the anticipated demand for multi-purpose, multi-process equipment.

v

Equipment manufacturers are also anticipating a large increase in the demand for ion process equipment over the next decade. It is clear that the United States and Canada have lagged behind Western Europe, Eastern Europe and Asia in utilization of plasma processing on a routine basis. The equipment manufacturers anticipate that North America will catch up and surpass their colleagues around the world during the next ten years

The viability of offering commercial ion nitriding services on a toll basis has been well established in Europe beginning in the mid-50's. North American heat treat services are now just recognizing the fact that these processes represent major revenue opportunities and are turning their attention to providing these services for their clients. The panel discussion of commercial shops offering daily ion nitriding and ion carburizing services proved to skeptical potential users that the process merits their attention and, while requiring techniques different from traditional atmosphere and vacuum case hardening techniques, requires less "art" then has been developed for conventional processes such as atmosphere carburizing. Certain large North American captive heat treat shops are now growing at such a rate in plasma processing equipment, that they are manufacturing their own equipment in order to keep pace with the demand for their services. This was a pattern that was established in Europe (particularly in Germany) and again, the North American continent seems to be lagging behind its European and Asian colleagues. The concern is that this gap will be closed over the next decade. It is estimated that there are 1,000 ion nitriders operating in the Peoples Republic of China, as well as 30 ion nitriders surrounding Paris, while only an estimated 100 ion nitriders are in service in North America. These numbers indicate that a large potential for the process exists worldwide which should and will be felt shortly in North America.

Ion nitriding, as a process, is proving to have a diversity of uses. Research on new applications continues unabated. While the basic discovery of the process occurred in the late 20's and early 30's, the application to new materials and multi-step processes continues along with the continuation of investigation of process parameters, gas mixtures to generate unique properties. The largest single unsolved problem which could create an enormous demand for ion nitriding as a process is the fact that ion nitriding of aluminum is theoretically possible, yet remains an enigma in practical application. The potential for nitriding of aluminum, niobium and other special alloys also represents an area of intense interest due to the process parameters and properties that can be achieved.

The adaption of the process by large automotive companies and automotive suppliers indicates that plasma processes will be one of the high production, diffused case hardening processes in the future. The use of robot handlers for loading/unloading and computers allows fully automatic processing of automotive parts production. Reports of parts processed to 90' long weighing 30 tons, as well as captive shops reporting 20-30 tons per weeks of ion nitriding to all service work going through their shops, to applications as small as ball bearings and ballpoint pen tips, indicate the total diversity of the process in production applications.

It is also interesting that ion carburizing seems to be going directly from bench scale, pilot plant directly to automotive part production on a heavy volume basis. New equipment designed to meet the demand for automotive type part production, along with research on the applications of plasma processes in multi step processes and to other materials indicates that the next conference will hold as much interest and attention as the past two have had and will establish the plasma diffused case processes as the most existing, commercial viable advanced case hardening processes for the next decade.

W. Kovacs and T. Spalvins
Conference Co-Chairmen

TABLE OF CONTENTS

MECHANICAL AND TRIBOLOGICAL PROPERTIES

INDUSTRIAL APPLICATIONS AND PERFORMANCE

PROCESSING AND PARAMETERS, DESIGN AND PRODUCTION EQUIPMENT

ION CARBURIZING

ADVANCES AND DIRECTIONS OF ION NITRIDING/CARBURIZING

Talivaldis Spalvins
National Aeronautics and Space Administration
Lewis Research Center
Cleveland, Ohio, 44135, USA

ABSTRACT

Ion nitriding and carburizing are plasma acti-vated thermodynamic processes for the production of case hardened surface layers not only for fer-rous materials, but also for an increasing number of nonferrous metals. When the treatment varia-bles are properly controlled, the use of nitroge-nous or carbonaceous glow discharge medium offers great flexibility in tailoring surface/near-surface properties independently of the bulk properties. The ion nitriding process has reached a high level of maturity and has gained wide industrial acceptance, while the more recently introduced ion carburizing process is rapidly gaining industrial acceptance.

The current status of plasma mass transfer mechanisms into the surface regarding the forma-tion of compound and diffusion layers in ion nitriding and carbon build-up ion carburizing will be reviewed. In addition, the recent devel-opments in design and construction of advanced equipment for obtaining optimized and controlled case/core properties will be summarized. Also, new developments and trends such as duplex plasma treatments and alternatives to dc diode nitriding will be highlighted.

THE CASE HARDENING PROCESSES such as nitriding, carburizing and carbonitriding are well estab-lished thermochemical diffusion processes for surface/near-surface hardening of ferrous materials, mainly steel. These processes are widely used in the manufacturing and machining industries primarily to treat engine components, tools for hot/cold work, die castings, machine tools, and tribocomponents. These processes are based on diffusion of the nonmetallic intersti-tial elements, carbon, and/or nitrogen, into the surfaces. Hard surface compound nitride or car-bide layers with an extended diffusion zone are formed. Nitrogen diffusion in steels is usually performed in the ferrite phase between 450 to 590 °C and is a ferritic treatment, whereas car-bon diffusion is performed in the austenite phase between 925 to 1050 °C and is an austenitic treatment.

To obtain the best response to nitriding steels, they should contain nitride forming ele-ments such as aluminum, chromium, titanium, molybdenum, vanadium, and tungsten. In carburiz-ing the maximum amount of carbon that can be dis-solved in the austenite phase can be determined from the $Fe-Fe_3C$ phase diagram. The nitriding and carburizing processes can be divided into two broad categories: conventional and ion assisted (plasma) processes. The conventional case har-dening processes are performed in solid, liquid or gaseous mass transfer media. These processes have been extensively described in the literature and will not be discussed in this review.

The ion nitriding/carburizing processes are plasma activated thermodynamic processes for the production of case hardened surface layers not only for ferrous metals but also for an increas-ing number of nonferrous metals. The ion assisted diffusion processes are based on the energetic nature of the glow discharge. The high energy of the nitrogenous or carbonaceous plasma contains ions, electrons, radicals, and activated species. The interaction between the plasma and the solid surface is based on excita-tion, ionization, dissociation and acceleration. The use of the glow discharge offers great flexi-bility in tailoring surface/near-surface proper-ties independent of the bulk properties when the treatment variables are properly controlled.

The ion nitriding process has reached a high level of maturity and perfection and has gained wide industrial acceptance primarily for increas-ing wear resistance and antigalling, and improv-ing fatigue life and corrosion resistance. However, the more recently introduced ion carbu-rizing process stands on the edge of industrial applications.

The objective of this introductory paper is to address the current status and understanding of plasma-mass transfer processes and plasma-surface interactions as to the formation of the compound layer and diffusion zone during case hardening. Further, the paper will highlight the recent accomplishments in equipment design/construction and the new trends in utilizing duplex plasma processes and the development of various improved/modified ion nitriding processes. Some general references (Refs. 1 to 5) on the subject described here are given in the references section.

ION ASSISTED SURFACE TREATMENTS

The ion assisted surface modification/deposition treatments can be classified in three categories:
  (1) **Ion assisted deposition** which covers physical vapor deposition (PVD) such as sputtering and ion plating, and chemical deposition (CVD) such as plasma enhanced deposition;
  (2) **Ion beam techniques** which cover ion implantation, ion beam mixing, and ion beam enhanced deposition; and
  (3) **Plasma thermochemical processes** such as ion nitriding, ion carburizing, ion carbonitriding, ion boriding, and ion oxidation.
Depending on deposition energies and surface interactions, the above processes can be classified as processes that produce distinct overlay coatings (ion assisted deposition) and processes forming no discrete coating but which modify the surface/near surface by diffusion, penetration and chemical reaction (plasma thermochemical processes, ion implantation). In this paper, the plasma thermochemical diffusion processes with regard to plasma-mass transfer/interactions will be described.

GLOW DISCHARGE CONSIDERATIONS

The glow discharge used as a processing plasma for surface modification can be established and sustained in various ways: dc diode discharges, rf discharges, microwave discharges, electron emission configurations and magnetically enhanced discharges. In the plasma assisted diffusion treatment, processes such as ion nitriding/carburizing, the dc diode discharge is most widely used. In these nonequilibrium plasma only the temperature of the electrons is high and that of the reactive species is relatively low. Because the temperature of the electrons is high, excitation and ionization of neutral atoms/molecules increase, resulting in enhanced reactive chemical reactions and plasma heating. Since the interactions of the energetic ion species with the surface are localized, the results which are obtained are different from that of simple (conventional) thermal activation.

Several types of glow discharges depending on the relationship between voltage and current exist as shown in Fig. 1. All plasma thermochemical processes are performed in the abnormal glow

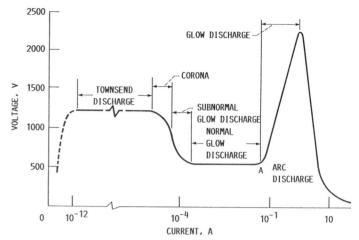

FIGURE 1. - VOLTAGE - CURRENT CHARACTERISTICS OF DIFFERENT TYPES OF DISCHARGE IN ARGON. (EDENHOFER) (REF. 6.)

in which the current increases with the voltage. In the abnormal region, the glow covers the specimen uniformly and therefore, uniform treatment can be expected. Further, the control of this unstable abnormal glow discharge is the critical feature in the production of uniform reproducible layers. With the advancement of technology, new plasma power generators have been introduced to prevent the formation of arcs, so that the glow discharge can be safely maintained at high current densities in the abnormal region.

The thermochemical processes such as ion nitriding are performed over a fairly well defined range of pressure and voltage (typically 1 to 10 torr and 0.3 to 0.8 kV) under dc diode plasma conditions. A typical ion nitriding system is schematically shown in Fig. 2.

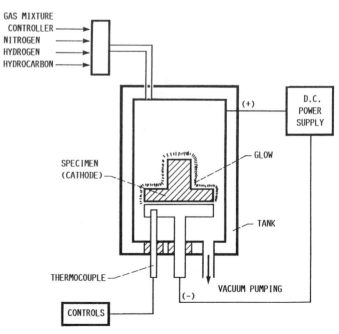

FIGURE 2. - ION NITRIDING SYSTEM.

ION NITRIDING/CARBURIZING MECHANISMS

Presently there is no universal model which explains the ion nitriding process. However, numerous mechanisms have been proposed to explain this concept. The fact remains that the exact reactant species in the glow discharge during ion nitriding are not well identified. As to ion carburizing, which is a more recent process, the basic explanations are derived from the more widely proposed ion nitriding concepts.

Basically, the ion nitriding/carburizing processes are controlled by plasma mass transfer (nitrogenous/carbonaceous species) and plasma-surface interactions. All the proposed mechanisms are essentially derived from these basic contributing factors:
(1) Sputtering,
(2) Ion excited atom implantation,
(3) Adsorption, and
(4) Condensation/deposition.
The numerous proposed mechanisms can be categorized and summarized as follows:
(1) Adsorption of atomic N;
(2) Sputtering: sputtered Fe and subsequent FeN condensation on surface (Kölbel's model) (Ref. 6), and sputter cleaned and activated surface; and
(3) Impact of molecular ions (NH^+, NH_2^+) and subsequent penetration.
In ion carburizing, the presently proposed mechanism is by C transfer from the CH_4/H_2 plasma to the specimen through the dissociation of CH_4 within the glow and forced attraction of the carbon species to the specimen (cathode) and finally reaction at the surface. All the above contributing factors are responsible for the enhanced plasma carburizing kinetics.

OPTIMIZATION OF PLASMA PROCESS VARIABLES

Up to now, only four process variables (temperature, composition of gas, pressure of gas and treatment time) have been controlled to optimize the microstructure, thickness, microhardness of the compound layer and diffusion zone. By varying the process variables such as composition of the gas mixture (nitrogen, hydrogen or hydrocarbon gases) one can custom tailor the metallurgical structures as shown in Fig. 3 to meet any specified requirements. The various nitride compound layers can be produced on the surface with a diffusion zone which may be 100 times thicker than the compound layer.

Recently, a fifth control variable, namely current density or power density can be regulated independently from the work-load temperature with the development of a plasma current density (pcd) sensor. In ion nitriding by controlling the power density/current density the compound layer thickness and composition can be accurately controlled, whereas during ion carburizing, the rate of carbon build-up on the surface can be monitored as shown in Fig. 4. Higher current densities increase the carbon build-up on the surface.

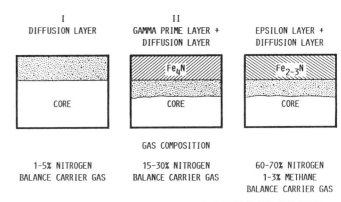

FIGURE 3. - METALLURGICAL CONFIGURATIONS DURING PLASMA NITRIDING.

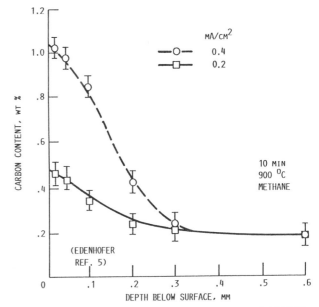

FIGURE 4. - EFFECT OF CURRENT DENSITY ON CARBON BUILD-UP DURING PLASMA CARBURIZING IN METHANE AT 900 °C.

RECENT DEVELOPMENTS AND FUTURE TRENDS

The current developments and improvements in the thermochemical diffusion treatment technology have been primarily in design/construction of more advanced equipment such as plasma power generators, and microprocessor controllers with computerization and programming. The objective is to optimize plasma and surface sensoring and to monitor/control all the process variables.

The future trends in the surfacing/modification technology are twofold: the development of duplex plasma treatments and alternatives to dc diode nitriding. The duplex plasma treatments are combined treatments where ion nitrided surface is subsequently coated by PVD (ion plating or sputtering) or by CVD coating techniques. Significant improvements have been achieved in the cutting tool industry where pre-nitrided drill bits are ion plated with TiN (Ref. 7). The numerous alternatives to dc diode nitriding which are presently investigated may be categorized as follows:

(1) Enhanced low pressure plasma nitriding which utilizes a dc triode configuration (Refs. 5 and 8),

(2) rf-nitriding which produces an enhanced discharge (Ref. 5),

(3) Pulsed-plasma nitriding at microsecond intervals for blind hole penetration (Ref. 5),

(4) Magnetically enhanced ion-nitriding for achieving deeper penetration (Ref. 5), and

(5) Plasma nitriding with air where gas mixture of air (nitrogen as donor) and H_2 (conc. critical) is used (Ref. 5).

Finally, the above techniques are extended to new materials such as powder metallurgy (P/M) products, powders and complex geometrical surfaces.

REFERENCES

1. Source Book on Nitriding, American Society for Metals, Metals Park, OH, 1977.

2. Ion Nitriding, T. Spalvins, ed., American Society for Metals International, Metals Park, OH, 1987.

3. Heat Treatment '84, The Metals Society, 1 Carlton House Terrace, London, England, 1984.

4. Plasma Reactions and Their Applications: Japan Materials Report, American Society for Metals International, Metals Park, OH, 1988.

5. Plasma Surface Engineering: Technological Trends and Impacts, (1st International Conference on Plasma Surface Engineering), A. Hauff, ed., Vol. 1, German Society for Metallurgy (DGM), Adenaueralles 21, D-6370 Oberursel bei Frankfurt/Main, Federal Republic of Germany, 1988.

6. "Physical and Metallurgical Aspects of Ion Nitriding. - I," B. Edenhofer, Heat Treatment of Metals, Vol. 1, 1974, pp. 23-28.

7. "Latest Work With TiN Coatings Extends Tool Life," A. Matthews and V. Murawa, Chartered Mechanical Engineer, Vol. 32, No. 10, Oct. 1985, pp. 31-34.

8. "Plasma Nitriding and Ion Plating With an Intensified Glow Discharge," A.S. Korhonen, E.H. Sirvio, and M.S. Sulonen, Thin Solid Films, Vol. 107, 1983, pp. 387-394.

COMMERCIAL AND ECONOMIC TRENDS
IN ION NITRIDING/CARBURIZING

William L. Kovacs
Elatec, Inc.
Andover, Massachusetts, 01810, USA

Looking at trends in the field of ion nitriding and ion carburizing, we have organized our approach to evaluate these trends:

The markets for plasma processes; the processes themselves; the equipment being offered; installation by captive shops as well as by commercial shops offering ion nitriding and ion carburizing services; and economic trends.

Data was collected by mailing over 2,000 survey forms worldwide. This survey was followed by a telephone and telefax survey to obtain specific information regarding advances in technology and process applications. One general comment can be made that, while the ion nitriding process has been known for about 60 years, it has consistently been reported at conferences and in literature as being a "new process". It would appear that this trend has ended and that ion nitriding, as a process, is no longer considered "new".

While metallurgists worldwide "know of" the ion nitriding process, it is clear from these surveys that, unless they own or are involved in ion nitriding itself, they have little actual information about the benefits in the process and how it differs from competing processes of salt nitriding, gas nitriding, fluid bed nitriding, or the coating processes such as titanium nitriding coating applied by chemical vapor deposition or physical vapor deposition, as well as the difference between ion nitriding and ion implantation of nitrogen. Recognition for the ion nitriding process has arrived, but differentiation of the ion nitriding process from conventional nitriding processes, as well as titanium nitride coating and nitrogen in implantation processes, simply does not exist. Sufficient ion nitriding installations and commercial applications exist, however, that metallurgists worldwide are able to recognize

that ion nitriding is commercially accepted and reliable.

It would appear that ion carburizing is benefiting from the 50 years of "advance publicity" provided by ion nitriding in that it is also "recognized" as a process. However, metallurgists outside the plasma processing field often put ion carburizing in the same state of development as ion nitriding, while people familiar with plasma processing recognize that ion carburizing is just beginning to realize its commercial potential. It would appear, however, that the process of ion carburizing is undergoing a distinctly different development trend than that of ion nitriding. Ion carburizing seems to be going directly from production size development equipment into production loads for large captive shops typically producing automotive parts. Typical examples of this development route are: Surface Combustion, Toledo, Ohio,[1] and Klockner Ionon Division, Leverkusen, West Germany.[2] In both cases, large production size, two chamber vacuum furnaces were used to develop the ion carburizing process and these furnaces were also used for prototype demonstrations of commercial automotive parts.

While we can count between 1300-1600 ion nitriding installations worldwide, its development and growth have definitely been through the classical route; "from laboratory bell jar, to bench scale, to "mini" works to commericial works". Further, this development has been undertaken by different companies and individuals worldwide, almost as if "they" have invented the ion nitriding process repeatedly, year after year. The development of ion carburizing, however, has been directly into production loads, as quickly as possible. The reason for this appears that the ion nitiriding process has more wide ranging applications to different materials and different metallurgical results that can be achieved by the ion nitriding process through control of the process variables, while the result desired from ion carburizing is primarily driven by desire to achieve a more uniform carburized case, with lower scrap, and better quality control.

We can report that the "markets for plasma processes" at this point seem to be opening very rapidly for the ion nitriding process, while the trend for ion carubirizng is to replace conventional carburizing processes through production of parts of high quality. Ion nitriding processes are now being combined with ion implantation, coatings applied by chemical vapor deposition, as well as physical vapor deposition. The driving force for these "multi-process" applications has been driven by the reports on enhanced wear by these dual treaments in cutting tools. In a study done for Westinghouse, Dr. Ram Kossowsky reported that tools that had been ion nitrided, followed by ion implantation of nitrogen, showed no measurable wear during tests.[3] Similar results are now being reported for wear parts, as well as tools using other coatings and implantation processes. Equipment designed for plasma assisted physical vapor deposition has always had the ability to also conduct the ion nitriding process. This equipment now appears to be used to apply both coating and ion nitriding treatments on cutting tools and other parts for enhanced wear. Companies such as Scientific Coatings and Hauzer offer these treatments as a commercial service.

If we look at the market for ion nitriding/carburizing, we see virtually unlimited potential and a market size in the "billion dollar class" for treatment services and equipment sales. While most people familiar with plasma processes recognize this enormous growth potential, the grim reality is that this potential is yet to be realized on a true commercial scale. The processes are just beginning their growth curve and the "time" when the period of rapid growth will begin still is uncertain. Most people surveyed "within" the plasma field believe this will happen within the next 3-5 years. In surveying commercial ion nitriding services who pioneered ion nitriding treatments on a toll basis in the United States eight to ten years ago (typically) now find these shops have installed 5-10 ion nitriding units. In itself, this represents remarkable growth in that the same phenomenon has not occurred with traditional process such as vacuum heat treating or coating processes such as TiN and TiC CVD and PVD coatings. All of the trends indicate that, between 1000-3000 ion nitriders should be installed in the United States and Canada over the next five to ten years. The potential for ion carburizing is even greater, but, recognizably, the timeframe will be longer.

When we look at trends, we see an exciting new market for the ion nitriding process through application to new materials. High temperature ion nitriding of refractory metals such as molybdenum and metals such as niobium and hafnium, as well as titanium, indicate exciting promise for the future. Perhaps the single largest commercial unrealized prospect is ion nitrding of aluminum. Theoretically, it is known to be possible and several researchers have reported successful results. Other researchers have been unable to duplicate these results, yet the prospect of putting a wear surface on conventional aluminum is a potential billion dollar market by itself. The carburizing process needs only to take its share of this market to achieve substantial commercial success.

The trend in equipment has also shown significant advances. After a shakey start due to the reliability of electronic controls in controlling the ion nitriding process, we can report that all the equipment being offered now is commercially reliable. One of the most exciting trends in equipment is the use of "multi-purpose" equipment. Elatec has delivered a true multipurpose system to a customer in Finland in which they plan to be able to sinter materials to 1650°C, and ion nitride these materials. The customer specified the equipment be supplied with a separate ceramic hot zone which can replace the graphite hot zone and allow the customer to use the furnace in an oxidizing medium for superconductor research and also for semiworks production of copper. Klockner reports a unique "multipurpose" furnace which has four chambers, which can conduct the process of ion nitriding and ion carburizing on a semicontinuous batch basis.[2] Elatec has also delivered a multipurpose unit to the New Jersey Institute of Technology which can be used to hot press, sinter, heat treat, as well as provide ion nitriding and ion carburizing treatments.

Another recent trend in equipment is innovations in fixturing. Equipment design now utilizes the fixture for the part as well as the capability of robotic loading to totally automate the loading and unloading of the parts.

Still another trend is the lowering of the cost of ion nitriding equipment. Many new companies have started producing ion nitriding equipment resulting in a drop in price on equipment offered to the customer. Additionally, it is now possible to retrofit any existing vacuum furnace into an ion nitrider or ion carburizer at a very low cost. This provides immediate multipurpose capabilities, as well as upgrades existing, surplus equipment to add these plasma process capabilities.

Virtually all equipment makers are reporting the ability to offer microprocessor control of the equipment. Microprocessor control allows true "hands-off" operation with no operator interface. While this is an important factor for automotive plants which require robotic loading and virtually no manual intervention, many commercial shops have reported going in the opposite direction whereby they report they have achieved better results by manual control of the process to force the operator to view the load during changes in pressure cycles. To service this need, equipment makers are offering control stations right at the viewpoints where the operator can directly "see" the effect manipulating the process variables has on the load.

One of the true controversies that exists among equipment makers is the battle over the use of pulsed high frequency D.C. supplies versus the conventional SCR fired D.C. supply. It would appear the controversy over the use of auxiliary heating in conjunction with the ion nitriding process has ended by the general acknowledgement that the use of auxiliary heating provides substantial benefits to the process and requires a minimum level operator skill when compared to D.C. only heated conventional SCR power units.

When we look at markets by key accounts, we see that the automotive companies, and parts suppliers to these companies, by far represent the largest commercial sector. In stating this, for ion nitriding, we are also recognizing that the plastics industry, from the molds through the actual machines, primarily services the automotive industry and that this industry perhaps has the single largest penetration of the ion nitriding process to every aspect of the industry. In respect to plastic molds themselves, ion nitriding is clearly replacing hard chrome plating on molds with high wear applications.

In comparing markets by geographic areas, we can see that ion nitriding is penetrating worldwide and ion carburizing has just a few installations. A survey of the 800 commercial shops in the United States and Canada indicates that 30% of these shops offer some type of nitriding service. Of these, 21% offer gas nitriding, 7% offer salt bed nitriding, 6% offer fluid bed nitriding, and 5% offer an ion nitriding service. Of the same 800 shops, 70% of these shops offer a carburizing service of which 48% offer gas carburizing, 19% offer "pack" carburizing, 12% offer salt carburizing, 5% offer fluid bed carburizing. Less than 2% offer vacuum carburizing, and less than 1% offer an ion carburizing service.

In conclusion, we can see that the commercial economic trends for ion nitriding/carburizing indicate growth in every area. All of the

markets appear to be growing and costs to be falling. If this trend continues, we would expect the ion nitriding/ion carburizing processes to emerge as the dominant case hardening processes over the next decade.

References

1. Personal communication with Dan Goodman, Surface Combustion, Toledo, Ohio.

2. Personal communications with Messrs Elwart, Remges and Hombeck, Klockner Ionon, Leverkusen, West Germany.

3. R. Kossowsky and R. E. Fromson, "Application of Ion Implantation to the Tool Industry", Proc. NAMRC-XII. Penn State University, PA, May 1984.

**Trends In
Ion Nitriding / Carburizing
By**

- Industries using the process
- Equipment
- New processes
- New materials
- Commercial shops offering plasma services
- Economics of the process(es)

FIGURE 1

**Trends Towards
"Key" Accounts For
Ion Nitriding / Carburizing**

Ion Nitriding	Ion Carburizing
Plastics industry (molds)	Automotive companies and their suppliers
Automotive companies (gears, valves, etc.)	

FIGURE 2

**Trends In
New Processes For Ion Nitriding**

Multiple Process (Ion Nitriding Plus)

- TiN Coating via CVD
- TiN Coating via PAPVD
- Ion Implantation (particularly N)

FIGURE 3

MARKETS BY MATERIALS

Ion Nitriding	Ion Carburizing
Ferrous Metals Refractory Metals	Ferrous Metals
Future Aluminum	

FIGURE 4

**Trends In
New Processes For Ion Carburizing**

- Rapid induction heated semi-continuous systems
- Multiple chamber units

FIGURE 5

**Trends In
Ion Nitriding Processing Equipment**

- Reliable equipment
- Multiple purpose furnaces
- Fixture design incorporated with automatic robotic loading
- Fully automatic microprocessor controlled systems
- Auxiliary heating on Ion Nitriding systems is becoming standard
- Equipment costs are lower
- Standard vacuum furnaces can be easily retrofitted to add plasma processing at a low cost

FIGURE 6

Equipment Manufacturers
USA - Equipment Manufacturers

Elatec	Mass
Abar/Ipsen	Penn
Seco warwick	Penn
VFS	Penn
Wellman	Ind
Beamalloy	Ohio
Consarc	NJ and Scotland
Surface Combustion	Ohio

Commercial shops making their own equipment

Sun Steel	Mich
Advanced Metal Technology	Texas
MPT	Cal/Texas/Germany

Canada	Germany	Japan
Iontec	Klockner	NDK
	Eltro	
	MPT	
	Plasma Technik	
	Surface Combustion	

France	China
BFI	Shanghi Electric Furnace Company

FIGURE 7

Trends In
Ion Nitriding / Carburizing Equipment Insatllations

Ion Nitriding 1300 to 1600

Ion Carburizing 20 to 50

FIGURE 8

Nitriding Services

(800 Commercial shops surveyed in the USA and Canada)

30% of these shops offer Nitriding services.
Of these:

- 21% offer Gas Nitriding
- 7% offer Salt Nitriding
- 6% offer Fluid Bed Nitriding
- 5% offer Ion Nitriding

FIGURE 9

Carburizing Services

(800 Commercial shops surveyed in the USA and Canada)

70% of these shops offer Carburizing services.
Of these:

- 48% offer Gas Carburizing
- 19% offer Pack Carburizing
- 12% offer Salt Carburizing
- 5% offer Fluid Bed Carburizing
- 2% offer Vacuum Carburizing
- 1% offer Ion Carburizing

FIGURE 10

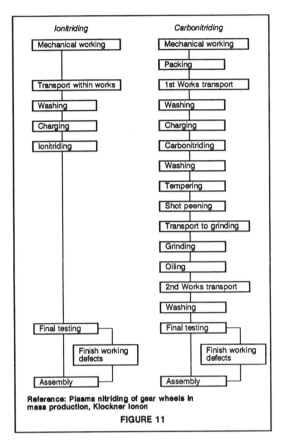

Reference: Plasma nitriding of gear wheels in mass production, Klockner Ionon

FIGURE 11

Economics
Assumptions

Equipment:	12 Furnaces for plasma treatment 9 Power units with control devices
Staff:	13 Persons, including profit centre head and secretary

Conclusions

Install	1	2	3	4	Commercial heat tr'm't shop
Profit before tax, %	14.2	0	22.2	39.5	19.4

Install	1	2	3	4	Commercial heat tr'm't shop
Return on invest-ment, %	4.6	0	9.8	40.5	9.7

Reference: Scientific and economic aspects of plasma nitriding, Klockner Ionon

FIGURE 12

FIGURE 13
Four chamber multipurpose Ion Nitrider/Carburizer
(courtesy Klockner Ionon).

FIGURE 15
Graphite resistance hot zone used in Multipurpose Sinter,
Hot Press and Ion Nitriding/Carburizing system at New Jer-
sey Institute of Technology.

FIGURE 14
Multipurpose Sintering, Heat Treating, Hot Press and Ion
Nitriding/Carburizing system installed at New Jersey
Institute of Technology (manufactured by Elatec, Inc.).

FIGURE 16
New concept in fixturing loads as part of furnace design
(courtesy Klockner Ionon).

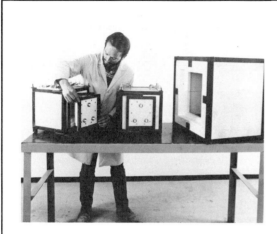

FIGURE 17
Atvac™ ceramic hot zones capable of being used with oxidizing or reducing gases in Elatec's Versivac™ and Aerovac™ Multipurpose furnaces.

FIGURE 19
Large Ion Nitriding chamber for automotive dies 36" x 36" x 96" (manufactured by Elatec, Inc.).

FIGURE 18
Retrofit of Brew Heat Treating furnace to add Ion Nitriding capability (conventional SCR fired D.C. power or pulsed high frequency D.C. power).

PLASMA DIAGNOSTICS FOR USE IN REACTIVE MAGNETRON SPUTTERING

S. M. Rossnagel
IBM
Yorktown Heights, New York, USA

ABSTRACT

A range of plasma diagnostic techniques have been applied to the study of reactive and non-reactive magnetron sputtering. Some of the diagnostics include probes, such as Langmuir, emissive and magnetic probes, optical emission spectroscopy, mass spectroscopy, energetic neutral and negative ion detectors, surface probes on both the cathode and the sample surfaces, neutral density probes, and energy analysis detectors. The general goal of plasma diagnosis for the field of reactive magnetron sputtering is the control of the deposition process, such that the cathode can operate at highest efficiency. Of the range of diagnostics, significant perturbations in the local gas density and temperature have been observed near the cathode. These effects can alter the chemistry at both the cathode and sample surface, depending on the type of chemical reaction. There is also an effect on the two main types of optical emission techniques, which can vary depending on the species and the type of experiment performed.

INTRODUCTION

Magnetron sputtering, and in particular, reactive magnetron sputtering have been in common usage for several decades for the production of thin films and hard coatings of various types. However, the level of understanding of the fundamental processes which occur both within the plasma as well as at the surfaces in contact with the plasma remains at an empirical, phenomenological level. A number of plasma diagnostics have been used to monitor the various conditions within the plasma. Some of these techniques include electrical probes, such as Langmuir or emissive devices, magnetic probes, optical emission spectroscopy, mass spectroscopy, energetic neutral or negative ion measurements, neutral gas density probes, probes built into the surface of the cathode or the sample, as well as energy analysis of the various energetic species.

Reactive magnetron sputtering is characterized by two regimes of operation. In the first mode, the cathode surface is in the metallic state, and the sputtering rate is characteristic of the original bulk metal. In the second regime, the cathode is covered by a compound layer, formed by a reaction with a gas species within the chamber. In this second mode, the sputtering rate is often significantly reduced from the metallic state, and as such, the deposition rate is reduced.

The state of the cathode is usually determined by the relative level of the reactive gas present, the operating power of the discharge and the pumping rate of the system. A common representation of this situation is shown in Fig. 1, which shows a hysterisis-like response of the deposition rate (or the discharge voltage) as a function of the amount of added reactive gas. Typically, the magnetron is sputtered in an inert gas plasma and the reactive gas is added to the chamber to produce a curve similar to Fig. 1.

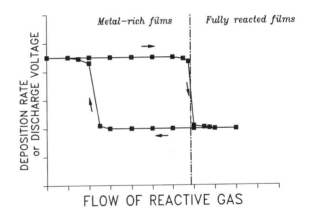

Figure 1. Typical deposition rate as a function of reactive gas flow for reactive sputtering showing hysterisis-like behavior.

In the first regime at low reactive gas flows, the cathode is being sputtered at high rate, but the deposited films are not fully reacted to the final compound state. This is due to a lack of the reactive gas species. As the flow is increased, eventually the films become fully reacted. However, this is an unstable operation point. Any slight increase in the level of the reactive gas species will result in the formation of too high a level of compounds on the cathode. If this happens, the sputtering rate of the cathode will drop and more of the reactive gas atoms will become available for deposition on the cathode, further reducing the sputtering rate. The system will quickly, and irreversibly transfer into the second regime. To return to the metallic mode, the level of reactive gas must be reduced to a very low level, such that the compound overlayer on the cathode can be removed by sputtering. At very low flows, then, the cathode can revert back to the metallic mode of operation.

Much of the work in reactive sputtering has been to monitor and control this transition from the metallic to the compound mode. The highest deposition rate, and the best films are found close to the knee of the curve. However, operation at this point can be complicated. One solution has been to increase the net pumping rate of the chamber many times. This results in such a large loss of reactive gas species (through pumping) that the chemical composition of the films on the walls of the chamber is no longer significant. While this approach has been successful for small, laboratory scale systems, it is not particularly practical for production devices.

The use of plasma diagnostics in reactive sputter deposition has been pointed directly at the control of the transition from the metallic to the compound modes. A number of the diagnostic techniques listed above have been used successfully (in some cases) to monitor the process and feedback-control some aspect of the system, typically either the discharge power or the flow rate of reactive gas.

DIAGNOSTIC TECHNIQUES

An extensive series of diagnostic measurements have been made on magnetron plasmas in both reactive and non-reactive modes. Langmuir probes (1-3) have been used to measure the electron density, temperature and plasma potential for a variety of conditions. In general, this type of probe is not used for reactive sputtering because the probe is rapidly coated with an insulating layer. Emissive probes have been used which measure the plasma potential by monitoring the emission from a hot filament (4). This type of measurement is compatible with reactive plasmas, but only measures the plasma potential, which is of marginal use in most cases. Magnetic probes have also been used to monitor the level of circulating current in a magnetron (5). This data is of information for understanding the transport processes of electrons, but has little use for reactive depositions.

Optical Emission Spectroscopy (OES) is often used to monitor the relative levels of various species within a plasma. The emission of light from the plasma can be exceedingly complex, but qualitative measurements are of great value for a technique such as end-point detection, where a reactive etching process is controlled by looking for emission from an underlayer as it is exposed. A related technique is stimulated optical emission, where gas atoms from the discharge chamber are excited by electrons or photons, and emit light which can be analyzed. This technique has two manifestations: the first is Laser-Induced Florescence (LIF) which is a valuable technique for monitoring species density as well as energy. A second device is conceptually much simpler, and is used for monitoring reactive depositions by monitoring emission from the reactive and non-reactive gas species. This second technique, as well as OES, will be discussed at more length below.

Mass spectroscopy has long been used to monitor deposition systems. Atoms from the discharge chamber are mass analyzed in semi-real time, and their density can be related to some aspect of the deposition process. This technique, however, tends to sample from the chamber as a whole rather than locally. As such, the response time is somewhat slower than the optical techniques, which can be more localized.

Another topic of interest during sputter deposition is the level of energetic, neutral species which bombard the sample during film growth. These species come from two sources: ions which are accelerated to the cathode, neutralized and elastically reflected from the surface, and negative ions which may form on the cathode surface during the bombardment, which are then accelerated across the sheath, stripped in the plasma, and impact the sample at high energy. This latter phenomena is quite relevant during reactive sputtering, and can in some cases lead to the removal of the film (6). The level of energetic neutral bombardment is difficult to calibrate because the detection of neutrals is complicated. Recent results have indicated a strong dependance on the relative masses of the ion and target atoms, with quite little pressure dependance in the range of interest to magnetron sputtering (7).

Another class of diagnostic techniques uses pressure sensors to sample local variations in either the direction or density of the gas atoms. First described as a "sputtering wind" (8), the turbulence and heating of the background gas atoms has been attributed to non-uniform heating by the somewhat energetic sputtered atoms from the cathode (9). The detection of these effects is straightforward, and can be shown schematically in Fig. 2. A small tube is positioned close to the cathode and is connected to a pressure transducer, which is typically a capacitance manometer. Changes in the gas density, pressure or temperature near the cathode can then be observed as a function of either chamber pressure or discharge parameters.

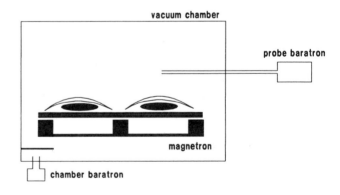

Figure 2. Sampling configuration for measuring local density variations during magnetron sputtering. The details of the

vacuum system and the requisite power supplies are not shown.

The results can be significant, as shown in Fig. 3 for the case of sputtering a Cu cathode in Ar. The density is strongly depleted in the near-cathode region as a function of increasing discharge current, or equivalently increasing the rate of emission of sputtered atoms.

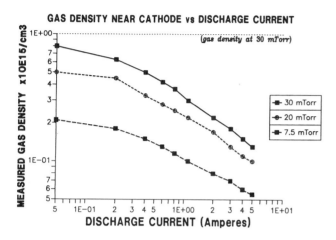

GAS DENSITY NEAR CATHODE vs DISCHARGE CURRENT

Figure 3. Neutral gas density in the near cathode region of a magnetron sputter deposition system as a function of discharge current. The cathode used was Cu and the gas was Ar.

This effect has been modeled successfully by calculating the energy deposited in the gas phase by the energetic sputtered atoms and the subsequent thermal conduction away to the relatively cold walls of the vacuum chamber (9). The model predicts an inverse-square relationship between neutral density and discharge current at high powers, which is confirmed experimentally in the data.

There are several implications of this type of result. The first is that the average gas temperature in these sputtering systems must be considered a strong function of discharge power, and that the temperature can easily reach into the $1000°K$ range. The second is that the local density is strongly perturbed as a function of discharge power, and therefore low power depositions will be more characteristic of high pressures, and high power depositions will be characteristic of low operating pressures.

The most significant implication from this type of result is that the chemistry at both the sample as well as the cathode will be strongly altered as a function of power. By driving the gas away from the near-cathode region as a result of this rarefaction effect, it might be expected that the formation of compounds on the cathode surface will also be affected, as will the formation of compounds on sample surfaces located close to the cathode (10,11). Unfortunately, conflicting results have been observed in this case.

As a function of discharge power, the number of sputtered atoms from the cathode increases approximately linearly. Therefore, the required amount of reactive gas to allow the deposition of the desired compound film should also increase linearly. If the gas is heated and driven away from the cathode-sample region by the result of the thermalization process, then additional amounts of the reactive gas will need to be added to compensate for the thermally-induced losses. Sproul and co-workers found this result for the sputtering of nitrides. However, work with depositing aluminum oxide by Rossnagel found just the opposite effect. As the discharge power was increased, proportionately less reactive gas was needed at higher powers. This result is shown in Fig. 4. It appears that a saturation-like effect is being observed in this case, which may be due more to the increase in oxygen atom temperature (velocity) rather than density. As nitriding reactions often require additional energy, whereas oxidizing reactions usually give off energy, it is possible that one process will be density driven, where the other will be temperature driven. It is often the case that oxide and nitride processes respond differently.

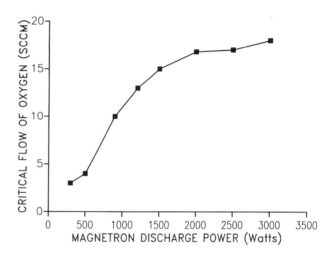

Figure 4. Required flow of reactive gas (oxygen) to cause the transition from metallic to oxide mode for an 200 mm. diameter Al cathode.

OPTICAL MEASUREMENTS

The two most commonly used optical measurements with reactive sputtering are Optical Emission Spectroscopy (OES) and a technique known commercially (12) as Optical Gas Controller (OGC). The first technique simply analyses the intensity and the wavelength of the light emitted from a processing plasma. The detector is often some sort on monochromator, configured with a grating or a prism, and a detector which is typically a photomultiplier or a diode array (Fig. 5a). The emission intensity can then be measured as a function of the wavelength in real-time. Any observed emission line in the plasma can then, in theory, be traced to an individual species, and the intensity of that emission line should be roughly related to the density of that species. The intensity of the line is proportional to the electron density in the plasma, the species density, and a complicated function of cross-section, excitation and de-excitation probability and electron temperature. For convenience, and typically because these parameters are not known particularly well in this type of plasma environment, this last function is assumed to vary only slowly over the range of interest. The electron density in the plasma can be approximately equated with the discharge current (13, 14). Therefore, the emission intensity of a species of fixed density in the plasma should roughly be proportional to the discharge current, which in a magnetron plasma is effectively equivalent to the discharge power.

This is indeed the case for the gas species at low discharge powers. However, as the discharge power is increased, the gas rarefaction effect begins to become significant, and the emission intensity changes to weaker levels (Fig. 6). As the local gas density is dependent on the inverse-square root of the discharge power, the emission intensity at high powers becomes proportional to the square-root of the discharge current.

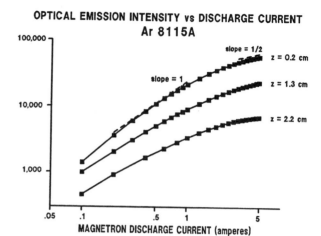

Figure 6. Emission intensity for Ar lines as a function of discharge current for the sputtering of a Cu cathode. The z-position is the vertical distance away from the cathode.

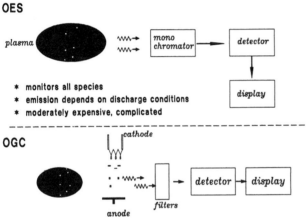

Figure 5. (a) Schematic of optical emission spectroscopy system. (b) Schematic of optical gas controller-type system.

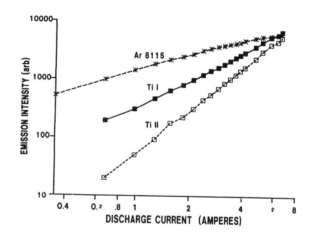

Figure 7. Emission intensity for Ar, Ti (Ti I) and Ti^+ (Ti II) as a function of discharge current.

The case for the sputtered metal species, however is significantly different. The neutral density of the sputtered species is directly related to the emission rate from the cathode, which is related to the ion bombardment rate or discharge current. Thus, the species density is proportional to the discharge current, as is the electron density. From this discussion, one would expect a quadratic dependance of the emission intensity for sputtered atom species as a function of discharge current. This effect is observed experimentally. A similar argument for ionized sputtered atoms suggests that the emission of these lines should be proportional to the cube of the discharge power. Indeed, this type of dependance is observed experimentally in Fig. 7 (14).

The second technique (OGC) relies on a non-plasma excitation process. A small filament is located in a chamber away from the plasma. The filament is heated, and electrons are drawn from the cathode to a separate anode. They are accelerated sufficiently that any impacts with gas atoms can lead to excitation of the atom, and subsequent photon emission. The photons are then filtered by narrow band filters and detected by a photomultiplier. It should be noted that it is also possible to use a monochromator at this same stage. By choosing the optical response of the filter, lines from individual dominant species can be identified.

This technique is quite similar to OES with one specific difference: the excitation function is effectively constant, and does not depend on plasma conditions, as is the case with

OES. Therefore, the emission intensity is a more clear indication of species density than is a signal at the same wavelength with OES. In addition, if metal species do enter with excitation region in the detector, they too will have a similar excitation, which is independent of the plasma properties.

The OGC technique has been successfully used by Sproul and co-workers to monitor the gas density near the cathode during reactive sputtering with nitrogen. A significant drop in the apparent pressure is observed, which is consistent with earlier studies showing a reduction in the local gas density (Fig. 8). The data of Sproul were not converted to densities. However, this can be easily done with the general assumption that an open vacuum chamber cannot sustain a pressure gradient of this type without a significant source of either flow or pumping. This is a similar assumption to the earlier work with manometer probes. It should be noted in the data of Sproul that the mass spectrometer traces in these figures show no dependance on the discharge power. This indicated that the overall chamber dynamics have not changed: only the volume close to the cathode. This, again, is quite consistent with the work with manometer probes.

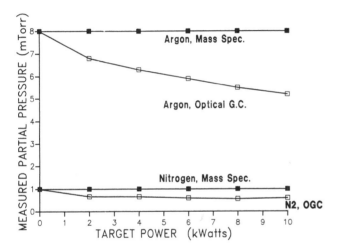

Figure 8. Measured apparent pressure with mass spectroscopy and the optical gas controller as a function of discharge power. Adapted from ref 10.

CONCLUSION

Plasma diagnostics can be exceedingly valuable in monitoring the state of a reactive sputtering system. Several of the results, however, suggest that the diagnostic techniques must be local to the cathode and sensitive to the perturbations in species density and temperature which take place there. Two very similar techniques, OES and OGS, operate in many ways under very similar principles. However, the linearity if the signal and the sensitivity to various species can be dramatically different.

REFERENCES

1. S.M. Rossnagel and H.R. Kaufman, J. Vac. Sci. & Technol., A4 1822 (1986).

2. T.E. Sheridan and J. Goree, J. Vac. Sci. & Technol., A7 1014 (1989).

3. B. Singh, presented at the American Vacuum Society National Symposium, Reno, NV 1984.

4. M.A. Lewis, D.A. Glicker and J. Jorne, J. Vac. Sci. & Technol., A7 1019 (1989).

5. S.M. Rossnagel and H.R. Kaufman, J. Vac. Sci. & Technol. A5 88 (1987).

6. S.M. Rossnagel and J.J. Cuomo, "AIP Conf. Proc" 165, 106 (1988).

7. S.M. Rossnagel, J. Vac. Sci. & Technol. A7 1025 (1989).

8. D.W. Hoffman, J. Vac. Sci. & Technol. A3 561 (1985).

9. S.M. Rossnagel, J. Vac. Sci. & Technol., A6 19 (1988).

10. W.D. Sproul, Thin Solid Films, 1989-in press.

11. S.M. Rossnagel, J. Vac. Sci. & Technol., A6 1821 (1988).

12. Inficon, Inc.

13. B. Chapman, "Glow Discharge Processes", Academic Press, New York, (1980).

14. S.M. Rossnagel and K.L. Saenger, J. Vac. Sci. & Technol., A7 968 (1989).

EXCITED STATES IN PLASMA NITRIDING

H. Malvos, C. Chave, A. Ricard
Lab de Physique des Gaz et des Plasmas
Bat 212, Univ. Paris-Sud
91405 Orsay, France

H. Michel, M. Gantois
Lab de Genie Metallurgique—Ecole des Mines
Parc de Saurupt
54042 Nancy, France

ABSTRACT

Excited states of nitrogen have been analysed by optical spectroscopy in flowing discharges and post-discharges for metal surface nitriding. Vibrational distributions have been determined for ionic and neutral nitrogen molecular states in glow discharges where the electron collisions are the dominant process and in afterglows which are produced by self-ionisation of $N_2(X,V)$ vibrationnally excited molecules and by N atom recombination. D. C. and H. F. glow discharges have been studied in a large range of gas pressures from 1-10 Torr for N_2 and 10-300 Torr for Ar-N_2 gas mixtures.

The vibrational temperatures of N_2 and N_2^+ excited states have been deduced from band head intensities. A discussion will be given on the excitation processes in discharge and post-discharge conditions.

Correlations are given between N_2 excited state production and nitrided layers growing on steel surfaces which are located in post-discharges.

INTRODUCTION

D. C. plasma reactors in N_2 and N_2-H_2 have been set up for surface nitriding of metals as steel to improve hardness and resistance to corrosion (1). In such reactors, the emission spectroscopy has been applied to characterize the ion and neutral active species in the negative glow near the steel substrate (2, 3). Then, post-discharge reactors have been setup in the purpose to separate the contribution of ions and of neutral active species in the steel nitriding process (4). By heating the steel surface (Fe-0.1%C) up to 840K and by applying a post-discharge treatment of two hours (N_2 discharge at 2 Torr, 80 mA, tube radius R=1cm), it has been obtained γ' Fe_4N layers of 6 microns depth with nitrogen in solution in α iron. nitrided layers were also elaborated with 2.45GHz microwave post-discharges (4). The main interest of H.F discharges is to operate without internal electrodes in a large range of gas pressure (10^{-4} Torr to atmospheric pressure) (5).

The vibrational distribution of positive and negative systems of the nitrogen molecule are presently described in discharge conditions where excitation is mainly produced by electron collisions and in post-discharges where reactions of molecular and atomic active species are the dominant kinetic processes.

N_2 DISCHARGES AND POST-DISCHARGES AT LOW GAS PRESSURES (1-10 Torr).

D. C. POSITIVE COLUMNS WITH INTERNAL ELECTRODES.

The D. C. plasma reactor in N_2 flowing gas in reproduced in Fig.1.

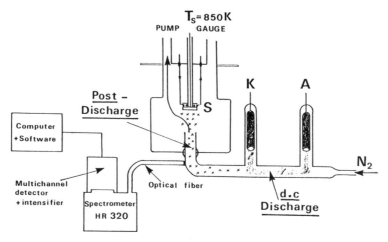

Fig. 1- D. C post-discharge reactor for steel surface nitriding. (S) substrate. (A) anode and (K) cathode. Spectroscopic analysis by the Jobin-Yvon MDS 320 system.

A.D.C discharge in N_2 flowing gas is produced upstream between two side-armed electrodes (cathode (K) - anode (A). nickel cylinders of dia. 0.8 cm) in a discharge tube of 2 cm in diameter. Downstream, the discharge tube is connected to a 15cm i,d glass cylinder where a XC 10 steel substrate (S) of dia. 3cm and knickness 0.8cm is located. The steel substrate is externally heated up to reach a surface temperature in the 800-850K range. After the N_2 d.c discharge, a post-discharge is characterized by an afterglow which has been analysed by emission spectroscopy. A Jobin-Yvon MDS 320 spectroscopic system including a 320 mm focal length monochromator with a 2400 gr.mm-1 grating and a multichannel detector (1024 pixels) with intensifier in the 200-920 nm spectral range has been connected by means of a quartz optical fiber with aperture matching optics to several parts of the N_2 discharge and post-discharge. The operating conditions of the N_2 discharge and post-discharge are a gas pressure in the 1-10 Torr range, a current discharge between 10-100mA, flow rates between 0.3-1.5 $N\ell$ min-1 giving a gaz velocity in the 10^2-10^3cm sec^{-1} range.

The emission spectra are characterized by positive and negative bands from excited neutral N_2^* and ionic N_2^{+*} molecules, respectively.

The N_2^* vibrational spectra between 570 and 610nm are reproduced in Fig.2a - for a 7 Torr-90mA discharge end (cf. Fig.1), 2b - for the maximum intensity in an early afterglow and 2c - for the late afterglow in the reactor outside the steel substrate. These spectra come from 1st positive emission N_2(B, V'-A, V") with V'-V"=4. The rotational structure of the vibrational bands is partly resolved with P_{11} and $P_{22}+^PQ_{22}$ branches as indicated in Fig.2a.

From the $I_{BA}(V', V")$ intensity of the P_1-heads, the [B,V'] vibrational densities have been calculated by the following equation:

$$I_{BA}(V',V") = C(\lambda) \frac{A(V'-V")}{\lambda} [B,V'] \quad (1)$$

where $C(\lambda)$ is a factor related to the spectral response of the detection system, λ is the optical wavelength and $A(V',V")$ is the radiative emission probability.

The vibrational distributions of the N_2(B,V') excited state are reported in Fig.3 for the discharge and the early and late afterglows in the post-discharge. A Boltzman distribution has been found in discharge conditions, characterizing a T_V^B vibrational temperature, according to the equation:

$$\frac{[B,V']}{[B,0]} = \exp - \frac{hc\, G_B(V')}{kT_V^B}$$

$$(2)$$

$$\text{with} \quad \frac{hc\, G_B(V')}{k} = 21.840K$$

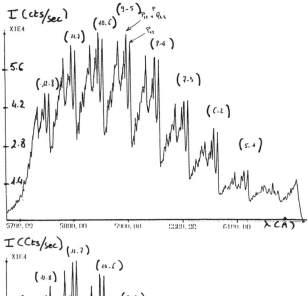

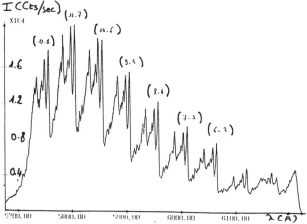

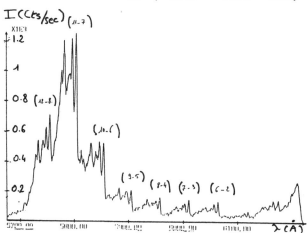

Fig. 2 -Vibrational spectra of N_2^* between 570 and 610nm. Resolution limit 0.2nm.

2a - 7Torr, 90mA discharge end with a 50msec.residence time.

2b - Maximum intensity in an early afterglow located between 25 and 90 msec after the discharge end.

2c - Late afterglow in the reactor outside the steel substrate, at time 0.2 sec.

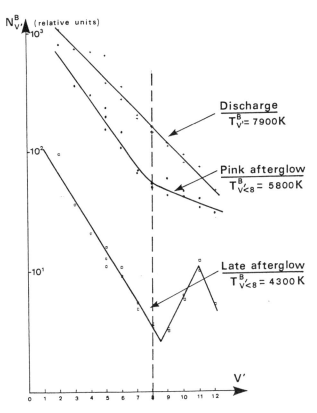

Fig. 3 - Vibrational distributions of $N_2(B, V')$ densities for $2 \leqslant V' \leqslant 12$. Discharge at 7 Torr, 90mA.

In the condition of N_2 discharge at 7 Torr, 90mA and for a residence time of 5×10^{-2} sec, the $T_{V'}^B$ - vibrational temperature is 7900K.

In post-discharge regimes, a nearly Boltzman distribution has only been found for low vibrational levels $V' \leqslant 8$ of the $N_2(B)$ state.

The corresponding vibrational temperatures are $T_{V' \leqslant 8}^B$ =5800 K in the early afterglow and $T_{V' \leqslant 8}^B \simeq 4300$ K in the late afterglow. The high vibrational levels are out of equilibrium in the afterglows with a maximum for $N_2(B, V'=11)$ in the late afterglow. This maximum is the result of N-atom recombination following the reaction:

$$N + N + N_2 \rightarrow N_2(^5\Sigma) N_2$$

(a)

$$N_2(^5\Sigma) \rightleftarrows N_2(B, V' = 11),$$

where the $N_2(B, V' = 11)$ state is produced from the $N_2(^5\Sigma)$ potential curve crossing (6).

21

The N atom density has been measured in the late post-discharge by the NO titration method (4). By introducing NO in the post-discharge, the following reactions are occuring:

$$N + NO \longrightarrow N_2 + O \qquad (b)$$
$$N + O + N_2 \longrightarrow NO\ (B) + N_2 \qquad (c)$$
$$NO\ (B) \longrightarrow NO(X) + h\nu\ (NO\beta),$$

at low NO flow rates.

Then, when all N atoms have been destroyed by NO, it follows:

$$O + NO \longrightarrow NO_2^* \qquad (d)$$
$$NO_2^* \longrightarrow NO_2 + h\nu\ (continuum),$$

at high NO flow rates.

The colour change from NOβ to the NO2 - continuum is produced when the NO and N flow rates are in equal quantities. The N-atom densities have been found to increase from 10^{14} to 10^{15} cm^{-3} when the d.c current discharge was varying from I = 10 to 120 mA (R = 1cm) at p = 2.2 Torr.

The vibrational spectra of N_2^* between 364 and 380 nm and of N_2^{+*} between 450 and 470 nm are reproduced in Fig.4 and 5a for a 7 Torr, 90 mA discharge, respectively. The N_2^* spectrum of Fig.4 concerns the vibrational band emission of 2nd- positive system $N_2(C, V'-B,V'')$ with $V'-V'' = -2$. The N_2^{+*} spectra of Fig.5 come from vibrational sequence of 1st- negative $N_2^+(B, V' - X, V'')$ with $V'-V'' = -2$. The N_2^+, 1st - neg. system has been observed in the discharge (Fig.5a) and in the early afterglow (Fig.5b) but the N_2, 2nd- pos. emission which is intense in discharge conditions disappeared in the early afterglow.

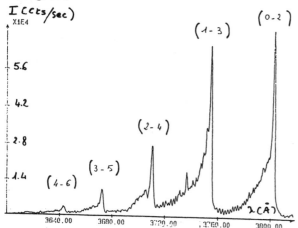

Fig. 4 - Vibrational spectrum of N_2^* between 364 and 380nm. Resolution limit 0.1nm. 7 Torr, 90mA discharge end.

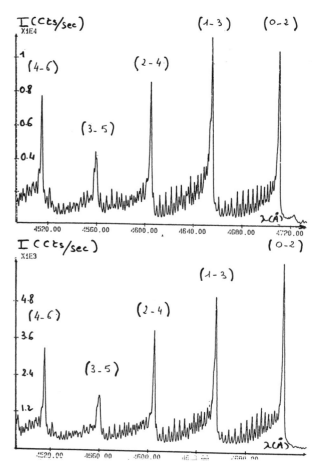

Fig. 5 - Vibrational spectra of N_2^{+*} between 450 and 470nm. Resolution limit 0.1nm. 5a - 7 Torr, 90mA discharge end; 5b - Early afterglow- Other conditions as in Fig.1a and 1b.

The early afterglow which has been designated as pink afterglow in previously work (7) is thus characterized by a strong $N_2^+(B)$ emission.

It has been previously suggested (7) that this $N_2^+(B)$ ionic emission is the result of $N_2^+(X)$ excitation by high N_2 vibrational levels, as given by the following reaction:

$$N_2(X,V +12)+ N_2^+(X) \longrightarrow N_2(X,V)+N_2^+(B,V'), \quad (e)$$

where the N_2^+ ions are produced in part by the following pooling reaction (8):

$$N_2(X,V \geq 32)+ N_2(X, W \geq 32) \longrightarrow N_2^+ + N_2 + e \quad (f)$$

The band heads in Fig.4 and 5 are unresolved rotational P- branches ($\Delta K = -1$) but the rotational R-branches ($\Delta K = +1$) which are degraded to the violet are nearly resolved. The vibrational distributions of $N_2(C,V)$ and $N_2^+(B,V')$ have been determined from the band head intensities of Fig. 4 and 5 by using an equation simular to eq.1. It has been found nearly Boltzman distributions of $N_2(C,V')$ and $N_2^+(B, V')$ vibrational states. The deduced vibrational temperatures of $N_2(C,B)$ and $N_2^+(B)$ states are reproduced in table 1 for discharge and post-discharge conditions at 7 Torr, 90mA.

Table 1 - Vibrational temperatures of $N_2^+(B,V')$, $N_2(C,V')$ and $N_2(B,V')$ states in N_2 discharge (residence time t_R) and post-discharge (Δt after the discharge end) conditions at 7 Torr, 90mA.

	Discharge		Pink afterglow	Late afterglow
T_V. $(10^3 K)$	$t_R \sim 10^{-3} - 10^{-2} sec$	$t_R = 5 \times 10^{-2} sec$	$\Delta t \sim 5 \times 10^{-2} sec$	$\Delta t \sim 0.2 sec$
1^{st}neg. $T_V^{B^+}$	5	5	4	
2^{nd}pos. $T_V^{C^+}$	5	5.5		
1^{nd}pos. T_V^B		7.9	5.8	4.3

Nearly identical values of vibrational temperatures are found for $N_2^+(B)$ and $N_2(C)$ which are lower ($\sim 5 \times 10^3 K$) than for $N_2(B)$ ($\sim 8 \times 10^3 K$). The vibrational temperatures are decreasing by about 20-30% when going from the discharge end to the pink afterglow.

MICROWAVE POST-DISCHARGES.

A microwave post-discharge reactor for metal surface nitriding is shown in Fig.6. The plasma has been initiated with microwave cavities, namely surfatron at 390MHz and surfaguide at 2450MHz (9), at gas pressures of 2-4 Torr and transmitted power varying from 100 watts at 390MHz to 800 watts at 2450MHz, in a quartz tube of 1.6cm int. dia. The plasma reactor of Fig.6 has been set up with a gas inlet pipe of 0.5cm int. dia. to work in a large range of gas pressure from about 3 Torr in N_2 to the atmospheric pressure in Ar-N_2 gas mixture.

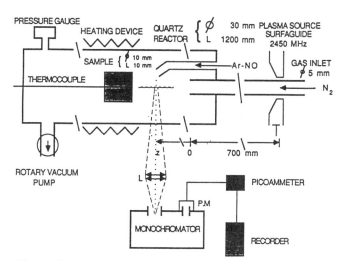

Fig. 6 - Microwave post-discharge reactor at low (1-10 Torr) and high (10-300 Torr) gas pressures for metal surface nitriding.

With the microwave plasmas at 2-4 Torr, it has been observed the early and late afterglows as for the D.C flowing plasma with internal electrodes. The N_2^+ 391.4nm intensities are reported in Fig.7 downstream the 390MHz surfatron cavity for 80, 120 and 180W transmitted powers. It can be observed in Fig.7 that the early afterglow maxima are produced at times of 11 msec (180W) and 13 msec(80W) after the plasma ends.

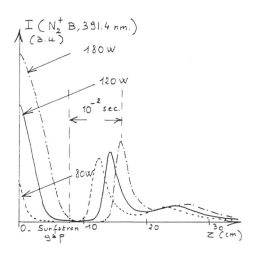

Fig. 7 - Intensities of $N_2^+, \lambda = 391.4$nm band downstream a 390MHz surfatron cavity. Discharge tube R=1.6cm, N_2 gas pressure 3 Torr, flowrate 0.4 Nℓmm^{-1}.

As for the d.c discharge and post-discharge, the vibrational distribution of $N_2(B,C)$ and $N_2^+(B)$ states have been determined and the vibrational temperatures have been deduced from the Boltzman graphs. The results are reproduced in table 2 for the 390MHz, 100 watts and the 2450MHz, 800 watts HF discharges and post-discharges.

Table 2 - Vibrational temperatures of $N_2^+(B,V')$ $N_2(C,V')$ and $N_2(B,V')$ states in N_2 390MHz, 100W and 2450MHz, 800W. H.F discharges and post-discharges at 3 Torr, 0,3-0,4Nℓmin-1.

$T_V(10^3)$	Discharge		Pink afterglow		Late afterglow	
	390MHz	2450MHz	390MHz	2450MHz	390MHz	2450MHz
1st neg. $T_{V'}^{B^+}$	8	13-20	4.5	7.5	7.5	
2nd pos. $T_{V'}^C$	12	15-20				
1nd pos. $T_{V'}^B$	7.8	12	6.8	10	5.6	

The extinction times between the discharge ends and the pink afterglows are 1.2×10^{-2}sec and 8.5×10^{-3}sec at 390MHz and 2450MHz, respectively. The late afterglow of 390MHz has been recorded at a time of 0.2sec. Note that the discharge temperature is as high than 1500K($\pm$ 200K) near the gap of the 2450MHz, 800 watts surfaguide plasma. The discharge temperature has been determined from rotational structures of 2nd positive and 1st negative system (2,3).

The discharge temperature is decreasing from the surfaguide gap (1500K) to the discharge end (900K), corresponding to a residence time of 6.5×10^{-3} sec. The N_2^+ rotational temperature has been found near 950K in the pink afterglow.

By comparing the D.C and H.F vibrational temperatures in table 1 and 2 it can be deduced that the N_2^* and N_2^{+*} vibrational states are more excited in H.F than in D.C. The $N_2(C,V')$ vibrational temperature is well related to the $N_2(X,V)$ ground state vibrational temperature since a direct electron excitation following the reaction:

$$e + N_2(X,V) \longrightarrow e + N_2(C,V'), \qquad (g)$$

has been found in the negative glow near the cathode (2).

Consequently, the $N_2(X,V)$ vibrational states must be more populated in the 390MHz and 2450MHz H.F plasmas to produce excited $N_2^+(B,V')$ ions by the pooling reactions e) and f). Also, the extinction time is shorter in H.F ($0.8-1.2 \times 10^{-2}$sec) than in D.C (5×10^{-2} sec) post-discharges..

With the 2450 MHz surfaguide plasma, at about 100watts in the gas inlet pipe of 0.5cm. int. dia., nitriding treatments of Fe-0,1% C samples (cylinders of 10mm. in. dia. and 10mm in length) have been performed in the post-discharge reactor of Fig.6.

In order to compare the microwave post-discharge reactor (Fig.6) with the one of internal electrode system (Fig.1), the nitriding treatments have been performed with about the same total gas pressures of 2-2.3 Torr (3mbar) and flow rates of 0.3-0.4Nℓ min.-1 with pure N2 and N2-H2 mixtures.

At this low gas pressure, the reactivity of the 2.450MHzpost-discharge appeared to be optimal at distance z = 20cm, for a transmitted microwave power of 80Watts and a nitrogen flow rate of 0.3Nℓmin.-1. Initially introduced to protect the samples against oxidizing, a low (2-3%) H2 rate can be maintained during the treatment.

The designed microwave post-discharge has produced at 840K and in one hour iron nitrides layers of 6μ thickness. It appeared that similar nitrided lavers were elaborated with the two post-discharges, but with a treatment time shorter in H.F (one hour) than in D.C (two hours).

POST DISCHARGES IN AR-N$_2$ GAS MIXTURES AT HIGH PRESSURES (10-300 Torr)

A large yield of nitrogen atoms has been achieved from an Ar - N$_2$microwave discharge at high pressures from 10 Torr to atmospheric gas pressure, using two to three times less microwave power than with a pure N2 discharge (10). In such experiments a surfatron exciter (5a) was used to sustain an Ar-N2 discharge in a 4mm i.d, 6 mm O.d. quartz tube.

At atmospheric gas pressure, the plasma visible glow was made up of stable filaments (typically two) extending over about 5cm with an Ar - 0.5%N2 gas mixture, a flow rate of 10Nℓm-1 and 200 Watts in the 2450MHz surfatrons exciter. Downstream the plasma filaments, a visible afterglow was produced by the N + N + Ar recombination (10). This post-discharge extended over 50cm in the quartz tube and into ambient air as a 2 to 5cm long conical tip. By NO titration in the post discharge using an Ar-1.3%NO mixture, it has been determined a N2 dissociation yield equal to 22% and 5.6% for Ar-xN2 gas mixtures with x = 0.3% and 3%, respectively. The microwave power investigated ranged from 100 to 250 watts.

The emission spectrum between 570 and 610nm is reproduced in Fig.8 for a Ar-4% N_2 post-discharge at p=90 Torr, Q=3Nℓmin-1 following a 910MHz, 100watts microwave plasma (11).

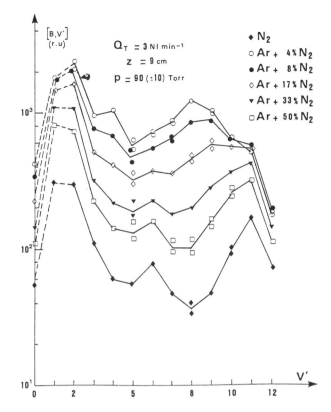

Fig. 9 - Vibrational distributions of $N_2(B,V')$ in 910MHz surfatron post-discharges for Ar-xN_2 with x = 4 to 100%.

The $N_2(B,V' = 11)$ peak in pure N_2 is progressively decreasing in front of a smooth maximum at $N_2(B,V' = 8)$ for Ar - xN_2 with x < 4%. By comparing Fig.9 and Fig.3, it can be observed a markedly non Boltzman distribution for $N_2(B,V'<8)$ in the N_2 microwave afterglow with a secondary peak at $N_2(B,V'=6)$. Also the $T_{V'}^{\ominus}$ temperatures reported in table 1 and 2 for the late afterglows must be only considered as apparent values. It is emphasized in ref.11 that the strong increasing of the $N_2(B,V' = 8)$ population when going from pure N_2 to Ar-4% N_2 post-discharges comes from a weaker quenching of the $N_2(B,V',8)$ state by Ar, following the reaction:

$$N_2(B,V'=8) + M_2 \longrightarrow \text{products} \qquad (h)$$

and not from an higher recombination rate following:

$$N+N+M_2 \longrightarrow N_2(B,V'=8) + M_2, \qquad (i)$$

with M_2 = Ar + x N_2.

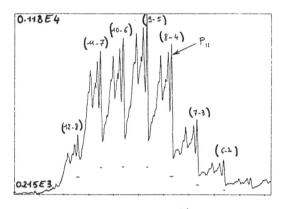

Fig . 8 - Spectrum of N_2 1st pos. in Ar - 4% N_2 post-discharge, Δv = 4 band sequence from λ = 575.5nm (12-8) to λ = 607nm (6-2) with a resolution limit δλ = 0.1nm. 910MHz surfatron discharge Q = 3Nℓ min⁻¹, p = 90 Torr, 100W.

By comparing Fig.8 with Fig.2c, it can be observed that the peak observed at B, V' = 11 in pure N_2 disappears in the Ar-4% N_2 post-discharge. The vibrational distributions of $N_2(B, V')$ deduced from eq.1 is shown in Fig.9 for the 910 MHz surfatron post-discharges at several Ar-x N_2 gas mixtures with x ranging from 4 to 100%.

By using the microwave post-discharge reactor as schematized in Fig.6 at high Ar-N$_2$ and Ar-N$_2$-H$_2$ gas pressures (50-500 Torr) and high flow rates (3-12Nℓ min-1), it has been observed that the 2450MHz surfaguide discharge is extending on few centimeters outside the exciter for a transmitted power of 70-80watts. Downstream a post-discharge filled up the reactor, surrounding the Fe-0,1%Csamples. After post-discharge treatments of 1 hour at 840K, it has been obtained γ' layer thichness of 4μm in Ar-1.4% N$_2$ (Fig.10) and of 2μm in A$_r$-1.4% N$_2$ 0.2% H$_2$(Fig.11).

The γ' layer thickness has been determined from the (110)α and (200)γ' intensity ratios of the diffraction patterns of Fig.10 and 11(4).

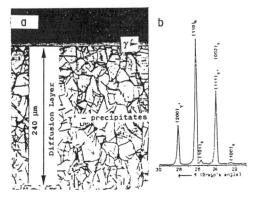

Fig. 10 - a) Cross section of a Fe -0,1%C sample nitrided 1h at 840K. Ar-1,4% N$_2$ post-discharge: 50 Torr, 72 Watts, 5.7Nℓmn-1. b) Surface X-ray diffraction pattern. γ': Fe$_4$N, ϵ: Fe$_{2-3}$N, α: α-ion.γ' thickness: 4μm, ϵ thickness < 1μm.

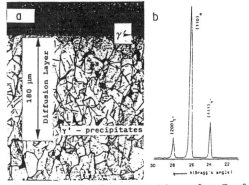

Fig. 11 - a) Cross section of a Fe-0.1% C sample nitrided 1h at 840K. Ar-1,4% N$_2$-0,2% H$_2$ post-discharge: 50 Torr, 72 Watts, 5.7Nℓ mn^{-1}. b) Surface X-ray diffraction pattern (λ K α_1 Co = 0.1789 nm). γ': Fe$_4$N,α: α-iron. γ' thickness: 2μm.

The steel substrate is very sensitive to oxidizing in the temperature range of nitriding treatments (resulting in part from water impurity inside the reactor). Thin iron oxide layers, especially magnetite could prevent the nitriding reaction. In order to avoid it, a small amount of H$_2$ has been introduced in the Ar-N$_2$ gas mixture. After 15 minutes of treatment, the H$_2$ gas inlet can be turned off without perturbing the nitriding process and without any growing of oxide layers. Results are reported in table 3 for three Ar-1.4% N$_2$ (H$_2$turn off) and one Ar-1.4% N$_2$ - 0.2% H$_2$ gas mixture at a distance z = 12cm in the reactor (cf. Fig. 6).

Table 3 - Nitrogen atom production and γ' nitrided layer depth in high pressure microwave post-discharges. Distance z=12cm. Treatment of 1 hour.

P(Torr)	Q Nℓ min-1	W Watts	N/N$_2$ %	depth μ
Ar - 1.4% N$_2$				
50	5.7	72	12($\pm$2)	4
170	5.7	80	2.3($\pm$.3)	3
340	12	80	1.1($\pm$.2)	1
Ar-1.4% N$_2$-0.2%H$_2$				
50	5.7	72	2.2 ($\pm$ .3)	2

By increasing the gas pressure from 50 to 340 Torr, the γ' layer is lowered from 4 to 1μ think. By analysing the results reported in table 3, it appears that the γ' thickness is well correlated to the N/N$_2$ dissociation rate as measured by the NO titration method (10).

CONCLUDING REMARKS

The emission spectroscopy is an in situ diagnostics of plasma-surface interaction which iswell suitable for control and drive a given treatment process. For metal nitriding plasmas, the vibrational distributions of N$_2^*$ and N$_2^{*\text{*}}$ excited states have been characterized in discharges where the electron excitation is the dominant process and in post-discharges where ionization reactions of vibrational N$_2$(X,V) molecules (pink afterglow) and recombination of N atoms are producing well typed afterglows.

Quantitative results of N atom densities in post-discharges have been obtained by the NO titration method in a large range of gas pressure from 1 Torr to the atmospheric gas pressure in Ar-N_2 gas mixtures. After N atom titration by NO for a chosen experimental condition, the 1st positive emission intensity in the late afterglow can then monitor the N atom densities in the post-discharge reactor.

REFERENCES

(1) J.P. Lebrun, H. Michel and M. Gantois. Mem. Sci. Rev. Metallurgie 69 (1972) 727.

(2) L. Petitjean and A. Ricard. J. Phys. D17 (1984) 919.

(3) J.L. Marchand, H. Michel, M. Gantois and A. Ricard, 86 Int. Conf on Ion Nitriding (Cleveland 1986).

(4) A. Ricard, J. Oseguera, H. Michel and M. Gantois. P. S. E. Conf. Garmisch (1988).

(5) M. Moisan, Z. Zakrzewski. Radiative Processes in Discharge Plasmas, p 381, Ed. J. M. Proud and L.H. Luessen, Plenum Publ. Corp (1986).

(6) H. Partridge et al. J. Chem. Phys. 88 (1988) 3174.

(7) J. Anketell. Can. J. Phys. 55 (1977) 1134.

(8) M. Capitelli, C. Gorse and A. Ricard. J. Physique Lettres 43 (1982) L417.

(9) M. Moisan, Ph. Leprince, J. Marec, Z. Zakrzewski et al. IEE Conf. Pub. 143 (1976) 382.

(10) A. Ricard, A. Besner, J. Hubert and M. Moisan J. Phys. B21 (1988) L 579.

(11) A. Ricard, J. Tetreau and J. Hubert. ISPC. 9 Bari (1989).

ION NITRIDING AND NITROGEN ION IMPLANTATION: PROCESS CHARACTERISTICS AND COMPARISONS

Arnold H. Deutchman, Robert J. Partyka, Clifford Lewis
BeamAlloy Corporation
Dublin, Ohio, 43017, USA

ABSTRACT

Ion nitriding and nitrogen ion implantation, two distinctly different surface treatment techniques, are compared and contrasted. Each process produces a case hardened surface layer, however the the implementation of each process, the characteristics of the case layers produced, the metallurgical strengthening mechanisms generated, and the economics and end-use of each, are quite different. Ion nitriding is a more mature technology that has already gained wide acceptance as a technique for the surface hardening of a variety of industrial components on a mass-production basis. Nitrogen ion implantation is an emerging surface hardening process that can be used either by itself, or as an augmentation to ion nitriding, and shows much promise for the surface treatment of precision tools and components.

ION NITRIDING is a thermally driven, equilibrium, diffusion process that produces a relatively deep (5 to 15 mils), hardened, case layer. Nitrogen ion implantation is a non-thermal, non-equilibrium, physically driven, ballistic alloying process, which produces a relatively shallow (0.04 mil), extremely hard case layer. Ion nitriding is implemented at high temperatures in a glow discharge atmosphere, while nitrogen ion implantation is carried out at room temperature, at high vacuum, in a dedicated atomic particle accelerator. Case layer strengthening in ion nitrided surfaces is due primarily to formation of transition metal nitride precipitates, while strengthening in nitrogen ion implanted surfaces is due primarily to dislocation pinning.

PROCESS IMPLEMENTATION

Ion nitriding is a thermal nitriding process in which the nitrogen ions to be diffused into the surface of the component to be treated are produced in a plasma discharge. The heat that drives the diffusion is provided predominantly by heat generated as nitrogen ions in the plasma discharge strike the surface of the component to be ion nitrided. Auxiliary heating is included as required. The temperatures required for ion nitriding of ferrous metals and aluminum are typically 1,000 degrees Fahrenheit, and the temperatures required for ion nitriding of titanium are in the 2,000 degrees Fahrenheit range. Ion nitriding times are on the order of 10 to 30 hours. In contrast to conventional nitriding techniques such as salt bath and gas ntiriding, the crystal structure of the layers in ion nitrided components can be tightly controlled. This is achieved by varying the composition of the gases in the plasma discharge.

The basic elements of an ion nitriding system are diagrammed in Figure 1. The components to be ion nitrided are mounted on a support structure inside of a double-walled, water cooled, vacuum chamber connected to a vacuum pumping system and gas supply system. The support plate is connected to a high current, high voltage DC power supply. The support plate is held at a negative potential (cathode) and the vacuum chamber acts as the positively charged anode. When voltage is applied under proper gas pressure conditions the glow discharge is produced. The glow discharge is simply an electric current passing through an ionized gas mixture and behaves as a function of applied voltage. Applied voltages are generally in the 1000 volt range. The ability to provide uniform ion nitrided case layers is tied directly to controlling the uniformity of the potential drop at the surface of the component to be treated. The electric field uniformity at the surfaces of the parts being nitrided is determined by the geometry of the surface. Electric flux concentrations will be different on flat surfaces, part edges and borders, and holes. Adjustments in gas pressures, operating voltages, and current densities during the ion nitriding cycle can generally guarantee uniform ion nitrided case layer characteristics.

Nitrogen ion implantation is the process of introducing nitrogen atoms into the surface layer of a solid material by accelerating the atoms to very high energies (50 to 200 KEV) and allowing them to strike the surface of the solid material. The energetic nitrogen ions penetrate into the surface of the material to depths ranging from 0.01 to 1 micron, depending on the energy of the nitrogen atom, and create a thin alloyed surface layer. Since the nitrogen atoms are injected into the surface of the material mechanically, there is no need for the application of high temperatures to produce a thermal diffusion of the incident nitrogen atoms into the surface.

Nitrogen ion implantation (Figure 2) is implemented by forming a pure beam of nitrogen ions and then providing the ions with kinetic energy sufficient to penetrate the surface of interest. A small linear accelerator is used to generate the nitrogen ions, focus the ions into a beam, direct the beam at the surface to be treated, and accelerate the ions to high enough energies so that they can penetrate the target surface. Magnetic mass analysis of the nitrogen ion beam may or may not be used depending upon the type of accelerator used.

The characteristics of the nitrogen ion implantation process are defined by the process requirements seen with the use of particle accelerator technology. The process is by definition a high vacuum process with operating pressure requirements in the $10E(-06)$ Torr, or lower, range. High vacuum environments are necessary so that the ion beams can remain focused. Nitrogen ion implantation is by nature a line of sight process. Ion beams with well defined envelopes, in order to optimize retained dose and minimize surface sputtering effects, must be directed at the surfaces to be treated at normal or near normal incident angles. For uniform implantation of parts with irregular surfaces, the parts must be manipulated in the beam to provide optimum incident angles throughout the treatment cycle. This implies that ion implantation processing chambers must be built with the capability to rotate, angle, and move parts during the ion implantation process.

Since ion implantation is a physical injection process as opposed to a thermal difusion process or a chemical coating process, thermal energy is not required for surface alloy formation. Parts can be processed at temperatures that do not exceed room temperature. Energy densities of the ion beam can, however, be very high, and care must be taken so that components that have temperature sensitive properties are not heated above critical levels. In general, beam accelerating voltages are in the 100,000 volt range, and with beam currents of 5 to 10 milliamps, relatively high energy densities can be achieved. Temperature rises in ion implanted parts can achieve 1,000 degrees Fahrenheit relatively rapidly, and all parts that are ion implanted must be mounted to platen assemblies that are water cooled so that excess heat can be removed from the components during treatment. With proper balancing of delivered beam currents and water cooling rates, temperature rises in the ion implanted parts can be limited to 100 degrees Fahrenheit or less. Since processing temperatures can be maintained below 300 degrees Fahrenheit, components that are nitrogen ion implanted can be treated without any risk of dimensional distortions or bulk property changes. There is no need to regrind, re-heat treat, or straighten components after ion implantation.

The key process characteristics of both ion nitriding and nitrogen ion implantation are summarized in Table 1.

CASE LAYER CHARACTERISTICS

The case layers produced by ion nitriding and nitrogen ion implantation are very different in depth, structure, and composition. This is to be expected since ion nitriding is a thermal diffusion process, and nitrogen ion implantation is a physical injection process. Figure 3 diagrams the basic differences between the case layers produced in steel by each process.

The case layer produced by ion nitriding shows the typical profile exhibited by thermal diffusion processes. The concentration of nitrogen peaks at the surface, and then decreases exponentially as a function of depth into the surface. The maximum atomic percentage concentration of nitrogen that can be diffused into the layer is determined by the thermal solubility of nitrogen in the base material. In steel this is approximately 10 atomic percent. Typical case depths produced in steel by ion nitriding range between 5 and 16 mils.

The case layer produced by nitrogen ion implantation shows the typical profile exhibited by a kinetic energy driven ballistic alloying process. The peak concentration of nitrogen is not found at the surface but rather deeper within the implanted layer. The implanted nitrogen concentration profile rises from the surface as a funtion of depth into the layer to the peak and then decreases rapidly. Since nitrogen atoms are physically "injected" into, rather than thermally "dissolved" in, the steel surface, much higher nitrogen concentrations can be achieved. Steel layers with atomic percentage concentrations of nitrogen in the 30 - 50 atomic percent range can be produced. These high case layer concentrations of nitrogen can also be produced in aluminum and other metals which show low nitrogen solubility.

The case layer depth produced by nitrogen ion implantation is determined by the collisional processes that slow down the energetic nitrogen atoms as they strike and penetrate the implanted surface. In most metals the range of penetration of the energetic nitrogen ions is a micron or less under 100,000 volt acceleration potentials. Higher implantation voltages do not produce dramatic increases in penetrated depth because of the nature of the stopping processes. The increased kinetic energy does not help drive the atoms deeper into the surface but just appears as excess heat in the surface as the atoms slow down.

Even though the case layer depths produced by nitrogen ion implantation are much shallower than those produced by ion nitriding, the much higher atomic percentage concentrations of implanted nitrogen provide case layer hardnesses much higher than that achievable with ion nitriding. This is due to major differences in the type of strengthening mechanisms produced in an ion nitrided versus a nitrogen ion implanted surface.

STRENGTHENING MECHANISMS

The hardness of the case layers produced by ion nitriding is generated primarily by chemical strengthening mechanisms. Hard metallic nitride precipitates are formed with alloying elements in the base material including chromium, vanadium, aluminum, silicon, and molybdenum. In a fully hardened steel substrate in which the bulk hardness is in the Rockwell 59 to 61 range, case layer hardnesses in the 62 to 67 range can be produced. The concentration of the metallic nitride precipitates formed will be limited by the amount of nitrogen that can be thermally diffused into the case layer. Since nitrogen ion implantation produces layers with much higher concentrations of nitrogen, chemical strengthening mechanisms should be enhanced and the appearance of other strengthening modes can be expected.

A series of both chemical and physical

strengthening mechanisms contribute to hardening in nitrogen ion implanted case layers. The supersaturated atomic percentage concentration of nitrogen allows the formation of much higher concentrations of metallic nitride precipitates than that found in ion nitrided surfaces. The majority of implanted nitrogen atoms however locate interstitially and tend to "condense" in dislocations (Figure 4). These high density concentrations of nitrogen atoms act to pin the dislocations, thereby strengthening the metal in the implanted case layer.

The high energy injection of nitrogen atoms in the implanted case layer also has a series of mechanical effects which help strengthen the implanted layer. The high energy implantation of nitrogen atoms produces high levels of residual compressive stresses in the implanted zone (Figure 5). Equivalent to a "micro" shot peening effect, the compressive stresses reduce the tendency of surface cracks to open up under wear thereby strengthening the surface. High energy ion implantation of nitrogen also produces a mechanical mixing effect in the treated surface. This mechanical mixing (Figure 6) acts to refine both the grain and nitride precipitate structures of the metal in the treated region. The refined grain and precipitate structures act to further harden the implanted layer.

In a fully hardened steel substrate the combination of chemical and physical strengthening mechanisms act to harden the ion implanted case layer into the 80 - 90 Rockwell equivalent range. Since the nitrogen ion implanted case layers are very thin, their hardness must be measured using microhardness testing equipment. The key case layer properties and strengthening mechanisms characteristic of both ion nitriding and nitrogen ion implantation are summarized in Table 2.

APPLICATIONS

Ion nitriding is an accepted technique that has been used for the surface hardening of both ferrous and non-ferrous components on an industrial mass-production basis for over twenty years. Used primarily to enhance surface wear-resistance, ion nitriding can be used as a direct replacement for conventional gas and salt-bath nitriding treatments. A variety of production tooling such as stamping and forming dies, and plastic injection molds, are routinely ion nitrided to produce wear-resistant case layers that help increase the performance and wear life of the tool. An increasing number of engineered components including gears, valves, and crankshafts are also now being ion nitrided on a large volume production level.

Ion nitriding services are available from many heat treatment centers and a large number of dedicated ion nitriding service centers. An increasing number of manufacturers are integrating dedicated ion nitriding equipment into their production operations, and a substantial number of commerical ion nitriding equipment suppliers market ion nitriding suystems. The cost for ion nitriding treatment of components is in the 20 to 80 cents per pound range.

The nitrogen ion implantation process was introduced to the industrial marketplace fewer than 5 years ago. For most applications it cannot be used as a direct replacement for either conventional gas and salt bath nitriding, or ion nitriding, due primarily to the much shallower case layer produced. It can however be used as a substitute for, or synergistically with, conventional nitriding treatments and other hardfacing treatments like PVD/CVD hardcoating in certain applications. The key application areas for nitrogen ion implantation are those in which thermally induced distortions of bulk properties (hardness and dimensions), and surface finish cannot be tolerated. Additional applications include those in which wear is abrasive or adhesive and the wear surface is not loaded heavily. Examples include precision tooling used for stamping and plastic molding, precision gauges, precision cutting tooling, and powder compacting tools. Few non-tooling, engineered components have yet to appear in the marketplace that are routinely nitrogen ion implanted. One notable exception is a line of nitrogen ion implanted titanium alloy orthopedic hip replacements.

Nitrogen ion implantation surface treatment services are currently available from a small number of ion implantation service centers. Standard, commercial-grade equipment designed specifically for nitrogen ion implantation is not available. Most nitrogen ion implantation systems are either one-of-a-kind, custom-designed units, or modified semiconductor ion implanters. And due to the complex nature of the equipment required, nitrogen ion implantation is in general more expensive than ion nitriding.

CONCLUSIONS

Ion nitriding and nitrogen ion implantation are two distinctly different surface treatment techniques. They are each based on different metallurgical techniques, they each produce different alloy structures in the surfaces treated, they are each implemented with different processing hardware, and they each find different uses and applications. Ion nitriding today is an established surface treatment process that has been developing and penetrating the manufacturing marketplace for over 30 years. Nitrogen ion implantation today is an emerging surface treatment process which is in an early stage of development and industrial market penetration. In the future, ion nitriding will find increasing use in applications where engineered components must be surface hardened in large volumes, at low cost. Relatively low precision, large volume requirement automotive components and certain aircraft components are potential prime applications. Current and future applications for nitrogen ion implantation include high precision, relatively low volume, high prime cost, tools and components. Both ion nitriding and nitrogen ion implantation are today, and will continue in the future, advancing the state-of-the-art in surface treatment technologies.

GENERAL REFERENCES

1. T. Spalvins, ed., "Ion Nitriding," Proceedings of an International Conference on Ion Nitriding, September 1986, ASM International, Cleveland, Ohio, 1987.

2. R. Kossowsky, ed., "Surface Modification Engineering, Volume II, Technological Aspects, CRC Press, Boca Raton, Florida, 1989.

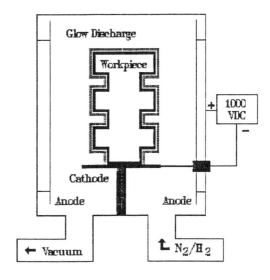

Figure 1. Basic elements of an ion nitriding system.

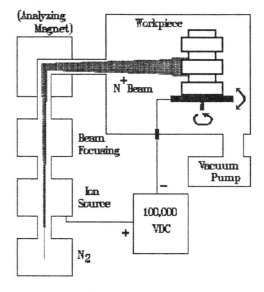

Figure 2. Basic elements of a nitrogen ion implantation system.

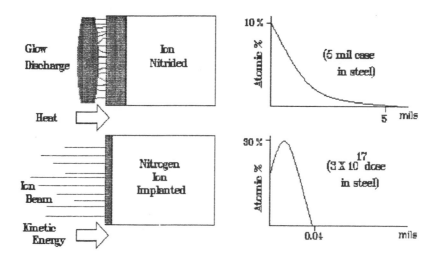

Figure 3. Case layer profiles for ion nitrided and nitrogen ion implanted layers.

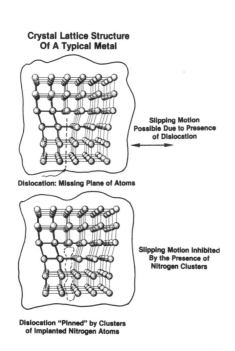

Figure 4. Case layer strengthening by dislocation pinning in a nitrogen ion implanted surface.

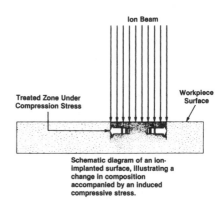

Figure 5. Case layer strengthening by compressive stress generation in a nitrogen ion implanted surface.

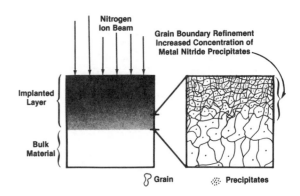

Figure 6. Case layer strengthening by grain boundary refinement in a nitrogen ion implanted surface.

Table 1 - Process Characteristics Comparison

Process	Type	Process Time (hr)	Process Temp.(F)	Process Pressure	Property Changes	
Ion Nitriding	Thermal Diffusion	10 - 30	900 - 2000	0.2 - 5.0 Torr	Hardness	Y
					Growth	Y
					Distort	Y
					Finish	Y
Nitrogen Ion Implantation	Physical Injection	1 - 6	< 300	10E(-06) Torr	Hardness	N
					Growth	N
					Distort	N
					Finish	N

Table 2 - Case Layer Properties Comparison

Process	Case Depth (in steel)	Atomic % Composition	Strengthening Mechanisms	Hardness (Rc equiv.)
Ion Nitriding	400 microns 16 mils 0.4 mm	10	Metallic Nitride	62 - 67
Nitrogen Ion Implantation	1 micron 0.04 mils 0.001 mm	50	Metallic Nitride Dislocation Pinning Compressive Stress	80 - 90

PLASMA SOURCE ION IMPLANTATION—NITRIDING CHARACTERISTICS OF STEELS

Murthy S. P. Madapura, John R. Conrad, R. Arthur Dodd, Frank J. Worzala
Department of Nuclear Engineering Physics
The University of Wisconsin
Madison, Wisconsin, USA

ABSTRACT

Conventional nitriding steel AISI 4340, martensitic stainless steel AISI 410, age hardenable LOW EX 43PH (invar type) steel, and TRIBOCOR (Nb base alloy) were nitrided by nitrogen implantation using Plasma Source Ion Implantation (PSII) and commercial Ion or Plasma Nitriding processes. The metallurgical characteristics and mechanical properties of the modified layers were studied to compare the two processes and to highlight the potential capabilities of Plasma Source Ion Implantation.

FOR ENHANCING WEAR, CORROSION, fatigue, and fracture resistance of engineering components, a variety of surface modification techniques are being used on a commercial scale. There is at present an increasing interest in ion or plasma nitriding and ion implantation treatment of surfaces due to their superior characteristics.

The literature on plasma nitriding and conventional beam-line ion implantation are quite vast. The characteristic process features, advantages, and limitations of conventional nitriding, plasma nitriding, and ion implantation nitriding are described in recent reviews [1,2]. In simple terms, ion nitriding is a thermochemical process operating in the glow discharge regime and the development of the modified layer follows the classical laws of thermodynamics and kinetics. It is an elevated temperature (400 - 550°C) non-line-of-sight process operating at relatively lower energies (0.3 - 1.0 KeV) and lower vacuum (1 to 10 torr). A variety of modified layer combinations can be produced, for example,

predominantly gamma prime or epsilon or a mixture of the two with a diffusion layer or only a diffusion layer. The property improvements of ion nitrided layers have been attributed to the type of compound layer formed and the fine dispersions of nitride precipitates in the diffusion layer.

In contrast, conventional ion implantation is a physical ballistic alloying process which violates the laws of thermodynamics and kinetics. It is an ambient or low temperature (generally less than 100°C), high energy (30 to 300 KeV), high vacuum (10^{-4} to 10^{-7} torr) line-of-sight process. These operating conditions result in a modified layer of very shallow depth (about one tenth of a micron). The improvements in properties have been attributed to the formation of very fine precipitates, high defect density, formation of amorphous and microcrystalline phases, and residual compressive stresses. The recently developed Plasma Source Ion Implantation (PSII) process is a non-line-of-sight process producing modified layers of equal or better characteristics compared to beam line implantation, without the necessity of beam stearing and target manipulation. The details of PSII are described elsewhere [3,4].

From the aforesaid description of ion nitriding and ion implantation processes and the characteristics of their modified layers, it could be inferred that both the processes are applied to engineering components for combating wear, friction, corrosion, and fatigue selectively. It is imperative that for comparing the two processes, development of equivalent modified layer depths is necessary. One way of achieving deep modified layers during ion implantation is through elevated temperature nitrogen implantation. The literature on elevated temperature ion implantation of steels is not extensive. Salik et al. [5], utilized a low energy nitrogen ion beam (0.5 to 1.5 KeV) instead of nitrogen plasma and obtained deep modified layers without compound layer formation in a relatively shorter time. However, they did not indicate the temperature experienced by the sample.

With this in view, elevated temperature nitrogen implantation by PSII was performed on several materials to develop thicker modified layers and to understand their metallurgical characteristics and tribological properties [6]. To facilitate their comparison with ion nitrided layers, samples having identical pre-implantation conditions were ion nitrided at a commercial facility. The results of this study should assist in comparing the characteristics of ion implantation nitrided layers with those formed by ion nitriding.

MATERIALS AND EXPERIMENTAL PROCEDURE

The materials used in this investigation are shown in Table 1. The samples were finished to less than $0.1\mu m$ surface finish by conventional grinding and polishing prior to treatment. The samples were implanted by the PSII and ion nitriding as per the process parameters shown in Tables 2 and 3 respectively. The as implanted/nitrided surfaces were characterized by profilometry, Knoop microhardness testing, X-ray diffraction, and pin on disc wear testing. Hardness depth profile, microstructure, and nitrogen penetration (by EPMA and SAM) were performed on cross-section samples.

Table 1 - Material and Condition

AISI 4340 - as quenched
AISI 410 - Quenched & Tempered
LOW EX 43PH - Solution Annealed
Tribocor (Nb-30Ti-20W) - Annealed

Table 2 - Implantation Parameters

Dosage: 5×10^{18} and 3×10^{18} atoms/cm^2
Temperature: 525 - 550°C on 4340 steel
  and 400 - 450°C on other samples
Energy: 53 KeV
Frequency: 165 Hz
Pulse Width: 10μsec
Current Density: 50μA/cm^2
Working Pressure: 4×10^{-4} torr
Time: 5 hrs. and 3 hrs.
Plasma Density: 3×10^9 nos/cm^3

RESULTS AND DISCUSSION

In surface modification of materials and components, the type of surface finish that can be achieved is very important both from the technical and esthetic view point. The surface

Table 3 - Ion Nitriding Parameters

Temperature: 550oC
Time: 5 hrs.
Pressure: 2.5 torr
Gas: 70H$_2$ - 30N$_2$

roughness profiles shown in Figure 1 reveal that implanted surfaces produced a variation of less than $0.3\mu m$, about 50% that of ion nitrided surfaces. The surface microhardness values shown in Table 4 reveal a significant increase due to nitrogen implantation in AISI 410, LOW EX 43PH, and Tribocor. In the case of AISI 4340 steel, the as quenched surface (starting condition) experienced simultaneous tempering, due to elevated temperature treatment, and hardening, due to nitriding. The competition between the two opposing processes resulted in a hardness decrease of about 16%, compared to the as quenched hardness. When compared with the tempered base material, the surface of implanted 4340 steel also showed a significantly higher hardness value. Appreciably higher surface hardness values at lower test loads in Tribocor revealed a shallower modified layer as the implantation temperature would not have exceeded 450°C which is considerably lower than 1900°C, the conventional nitriding temperature of this material.

A comparative assessment of surface microhardness of ion nitrided and implanted materials can be obtained from Table 5. It is evident from this table that surface hardnesses comparable to ion nitriding can be achieved by elevated temperature PSII nitrogen implantation. In the case of AISI 410 and LOW EX 43PH steels the substitutional elements Cr, Ti, and Al are expected to form fine nitride precipitates which impart high hardness. AISI 4340 steel, on the other hand, had a hardness which was 15% less than than the ion nitrided counterpart due to its low alloy content (about 1%Cr).

Evidence of a metallurgically controlled microstructure with elevated temperature nitrogen implantation is obvious from the X-ray diffractograms shown in Figure 2 and Table 6. This is significant for applications that require a

Table 4 - Surface Knoop Microhardness Before and After PSII- Nitriding

Material	Before PSII Nitriding (Test Loads - gf)			After PSII Nitriding (Test Loads - gf)			% Increase in Hardness (Test Loads - gf)		
	10	25	100	10	25	100	10	25	100
4340	750	730	690	690	630	580	-7	-14	- 16
SS410	490	410	320	1320	1120	990	170	170	200
LOW EX 43PH	400	340	310	1290	1220	860	230	260	180
TRIBOCOR	330	280	250*	2340	1230	670*	600	330	170*

* Measured at 50 gf

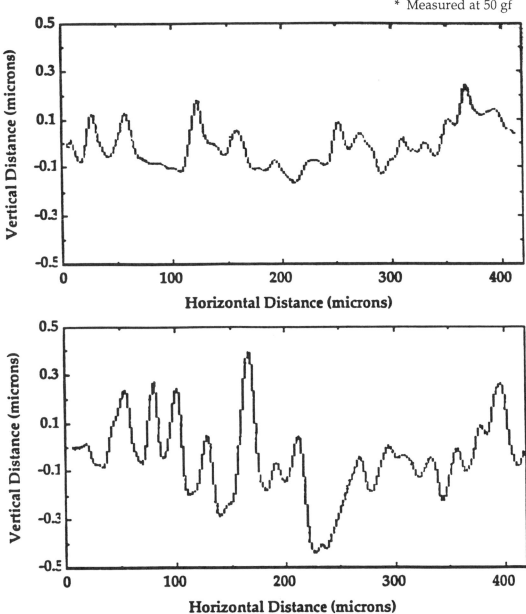

Figure 1 - Alphastep Profilometry Traces of Treated Surfaces in LOW EX 43PH Steel: Nitrogen Implantation by PSII, 3×10^{18} atoms/cm^2 (TOP) and Ion Nitriding (BOTTOM)

Table 5 - Surface Knoop Microhardness at 100 gf Load

Material	PSII - Nitrided	Ion Nitrided	%Variation w.r.t.IN
4340	580	688	- 15.7
SS410	993	1021	- 2.7
LOW EX 43PH	865	942	- 5.4

Table 6 - X-Ray Diffraction and Modified Layer Thickness@ from Knoop Hardness

Material	X-Ray Results		Modified Layer Depth (μm)	
	PSII-Nitriding	Ion Nitriding	PSII-Nitriding	Ion Nitriding
4340	No Compound Layer	Compound Layer	315	315
SS410	No Compound Layer	Compound Layer	30	75
LOW EX 43PH	No Compound Layer	Compound Layer	185	210

@ Hardness higher than 5% of the core value is the cut off point

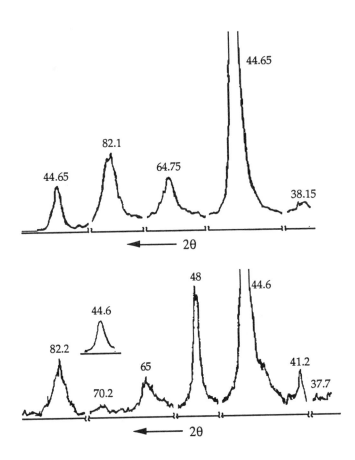

Figure 2 - X-ray Diffractograms of AISI 410 stainless steel: Implanted, 3×10^{18} atoms/cm^2 (TOP) and Ion Nitriding (BOTTOM)

4340 steel/Quenched and Implanted, 5X10^18

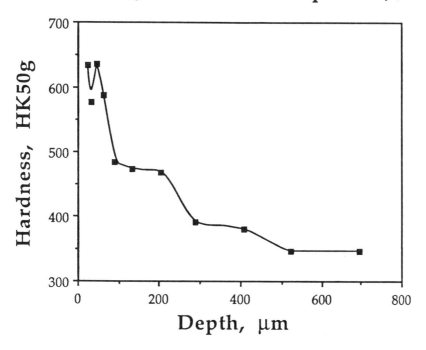

Figure 3 - Knoop microhardness - depth characteristics of nitrogen implanted AISI 4340 steel, 5X10^{18} atoms/cm^2.

Nitrogen Profile by Electron Probe Micro Analysis

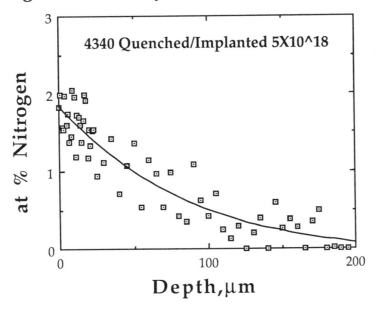

Figure 4 - Nitrogen concentration - depth profile - Electron Probe Microanalysis

high hardness but cannot tolerate the presence of a compound layer. Figure 3 shows the hardness profile of AISI 4340 steel which indicates the presence of a thick modified layer due to implantation. The nitrogen concentration profile shown in Figure 4 reveals a similar profile. Table 6 lists the modified layer depths observed from microhardness measurements on the cross-section samples. The layers obtained by implantation, even with low energy of nitrogen ions (53 KeV) and a low current density are comparable to those obtained by ion nitriding. The influence of fluence is shown in Figure 5. An increase of fluence, from 3×10^{18} to 5×10^{18} atoms/cm^2, shows a pronounced increase in the modified layer thickness as measured optically.

The photomicrographs shown in Figures 5, 6, and 7 clearly indicate that the modified layers achieved by elevated temperature implantation can be easily identified by optical metallography. However, to study the finer features, it will be necessary to perform transmission electron microscopy.

The summary of the results of pin on disc wear tests shown in Figure 8 reveal a remarkable improvement in sliding or adhesive wear resistance for both AISI 410 and LOW EX 43PH steels. The wear behavior of implanted and ion nitrided layers appear to show similar trends in these short term tests. The long term tests should show the differences if any in the wear and friction behavior of these layers.

SUMMARY AND CONCLUSION

1. The elevated temperature Plasma Source Ion Implantation process can produce thick, hard modified layers that are comparable to conventional ion nitrided layers.

2. The process produces a metallurgically superior modified layer without a compound layer.

3. The surface finish is significantly superior to ion nitriding.

4. The surface hardness and wear properties are comparable to ion nitrided layers.

5. The process can be used for a variety of materials that do not undergo any transformation in the temperature range and that require a harder and compound-layer-free surfaces.

6. By varying the process parameters, it may be possible to further reduce the processing time and temperature.

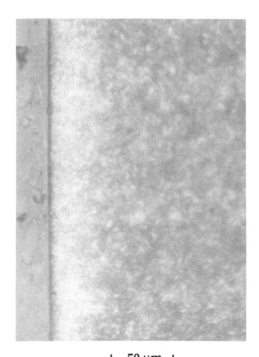

50 μm

Figure 5 - Optical micrographs of transverse sections of 4340/quenched and nitrogen implanted samples. 5×10^{18} atoms/cm^2 (TOP) and 3×10^{18} atoms/cm^2 (BOTTOM)

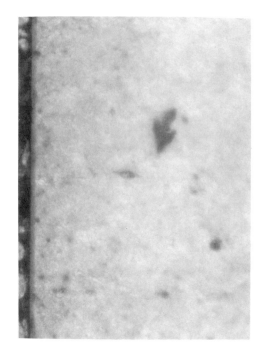

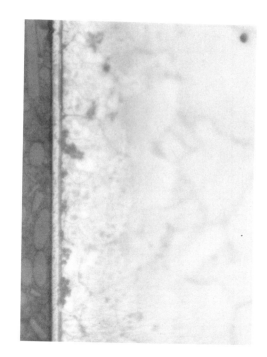

| 50 μm |

Figure 6 - Optical micrographs of transverse sections of AISI 410 stainless steel. Nitrogen implantation by PSII, $3X10^{18}$ atoms/cm^2 (TOP) and Ion Nitrided (BOTTOM)

Figure 7 - Optical micrographs of transverse sections of LOW EX 43PH steel. Nitrogen implantation by PSII, $3X10^{18}$ atoms/cm^2 (TOP) and Ion Nitrided (BOTTOM)

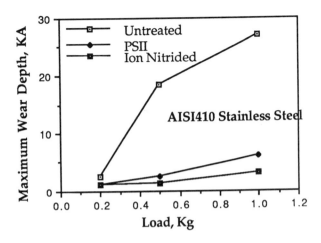

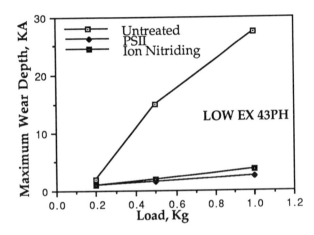

Figure 8 - Pin on disc wear characteristics of nitrogen implanted and ion nitrided samples: AISI 410 (TOP) and LOW EX 43PH (BOTTOM)

ACKNOWLEDGEMENTS

The authors acknowledge the technical assistance of P. Fetherston, J. Firmiss, and L. Xie in several phases of this investigation, and Prof. Everett Glover for his valuable help during Electron Probe Microanalysis. This work was supported by NSF grants DMC - 8712461 and CDR - 8721545, and by a number of industrial grants.

REFERENCES

1. dos Santos, C.A. and Baumvol., "Nitriding of Steels. Conventional and Ion Implantation", H. Ryssel and H. Glawisching, eds., Ion Implantation: Equipment and Techniques, Springer - Verlag, Berlin, (1983).
2. Hulett, D.M. and Taylor, M.A., "Ion Nitriding and Ion Implantation: A comparison, Metal Progress, Aug., (1985)
3. Conrad, J.R., Radtke, J.L., Dodd, R.A., and Worzala, F.J., " Plasma Source Ion Implantation Technique for Surface Modification of Materials", J. Appl. Phy., 62, 4591, (1987).
4. Conrad, J.R., "Method and Apparatus for Plasma Source Ion Implantation", US Patent, 4,764,394, Aug., (1988).
5. Salik, J. and Hubbell, T.E., "Ion-Beam Nitriding of Steels", US Patent, 4,704,168, Nov., (1987).
6. Worzala, F.J., Conrad, J.R., and Madapura, M.S.P., "Method for Surface Modification of Metals and Alloys for Producing Novel Microstructure and Improved Wear Resistance", Patent Disclosure, Submitted to WARF, (1989).

CURRENT STATUS OF PLASMA DIFFUSION TREATMENT TECHNIQUE AND TRENDS IN NEW APPLICATION

Kyong-Tschong Rie
Forschungsstelle für plasmatechnologische
Produktionsverfahren am Institut für Schweißtechnik
TU Braunschweig, Langer Kamp 8
D-3300 Braunschweig, FRG

1. Introduction

In view of the steady and expected expansion of plasma and ion assisted surface engineering into virtually every aspect of industrial activity, 2 years ago a status report /1/ was prepared by the initiation of the Federal Ministry of Research and Technology, Fed. Rep. of Germany. An extensive survey on entire field of plasma and ion assisted surface engineering has been made by scientists and engineers from various disciplines. The subjects concerned are summerized in table 1.

The objectives of the survey were

1. to present the state of the art,

2. to disclose the lack of knowledge,

3. to identify the trends in new applications,

4. to analyse and evaluate the current and future requirements,

Based on the report the directions of the future research as well as the recommendations for an intensive support of the research and development activities have been worked out and realized recently.

In the present paper a part of the above mentioned status report is reviewed. Only Plasma Diffusion Treatment (PDT) is dealt with in this lecture. Economic aspect of the PDT will not be included.

2. State of the Art

To meet the increased demand for better functional surface plasma and ion assisted technique has been developed to improve the surface properties. Recently this technique has been applied to synthesize new materials /2/.

Plasma Diffusion Treatment (PDT) is a most promising method to improve the surface and has been widely used over the past 20 years. One of the most striking advantages of PDT over conventional surface heat treatment is the wide scope of possible variation of process parameters.

Processes

DC glow discharge at relatively low pressure is mostly used for treatment applying several hundred volts for potential difference. Because of the high current densities requiered to cover the workpiece with "glow seam" which follows the contours of the workpiece, the glow discharge is in the region of the anomalous glow discharge which can be very unstable and easily transfered into an arc discharge. This has been the reason why the development of plant for commercial use was extremely sluggish although glow discharge techniques were first applied for surface engineering about fifty years ago. The development of thyristor controlled circiut breakers for arc suppression has led to a wide acceptance of PDT process and the arc suppression problem can be regarded as solved. Whether the use of the pulsed DC will be an additional help for arc suppression is still a matter of dispute. Unanimous advantage of the pulsed DC technique is however the possibility of the temperature variation without altering the plasma parameters such as voltages across the electrodes by simply changing the pulse duration and repitation time. It is ascertained that PDT facilities with pulsed DC power supply are increasingly offered by the equipment manufacturers, partly as an option.

Fundamental Studies

Because of the complexity of the species in plasma and of the inter-actions among species and species with solid surfaces it is impossible to present a unique picture of the mass transfer of interstitial alloying elements to the workpiece surface. Consequently the layer formation mechanisms are still not well under-stood. It has been often claimed that the following four surface reactions are involved:

1. Sputtering

2. Ionimplantation

3. Absorption of gas

4. Condensation/deposition

Several models have been proposed in the past to explain the mass trans-fer and layer formation.

A group of models is based on the mass transfer of neutral atoms to the workpiece surface. Tibbet /3/ has nitrided the iron and steel specimens in a holder provided with a grid to repel all positive ions from the spe-cimen surface. It has been shown that the nitriding rates for materials with or without the grid were clearly the same.

Based on the evidence achieved he claims that nitriding occurs primarily by the neutral atoms. This model was backed partly by the mass spectro-scopic investiations carried out by Szabo and Wilhelmi /4/. They found that the neutral particles of the form $FeNH_{2-3}$ are mainly responsible for nitriding.

Another group of models takes into account the mass transfer of ionized atoms and molecules. Lakhtin et al. /5/, Hudis /6/ and Johnes et al. /7/ claim that the mass transfer of the species NH^+ and NH_2^+ is a pre-dominent mechanism. For layer forma-tion, however, they claim that the ionized environment alone is not sufficient condition. The bombardment of the workpiece surface by partially dissociated species is necessary for nitriding. Employing RF plasma they observed that the nitriding occured only if potential difference across the electrodes was applied.

Kölbel suggested a model based on the sputtering by ion bombardment /8/. This model assumes that the condensa-tion of iron nitrides of the form FeN on the workpiece surface occurs through following four steps:

1. Ionization of nitrogen atoms

2. Sputtering of iron by ionized nitrogen

3. Formation of iron nitrides between sputtered iron and neutral nitrogen

4. Deposition of iron nitrides on the component surface

This model was again partially supported by the findings /9/ that a uniform iron nitride was formed even on nodular graphite of the cast iron after plasma nitriding.

Recently, Ricard et al. /10/ have performed a nitriding of steels in post discharge and obtained results which were comparable to the nitriding in glow discharge. He clearly demon-strated that the long lived active species are responsible for nitriding in post discharge. From their experi-ments it can be concluded that the sputtering is by no means a prere-quisit for plasma nitriding.

In the field of plasma carburiz-ing extensive spectrometric investiga-tions and modelling of plasma carbu-rizing have been performed by Dexter et al. /11/. They claim that excited neutrals and radicals are responsible for the enhanced carburizing in glow discharge.

Most of the plasma diagnostic investigations was carried out using RF, microwave plasma or DC glow dis-charge. Spectroscopic investigations applying pulsed DC in nitriding atmo-sphere were performed only by Szabo and Wilhelmi /4/, but the ratio of the pulse duration and pulse repitation time was not varied in their investi-gation.

Ferrous Materials

Ferrous materials are extensively nitrided and nitrocarburized in glow discharge in industrial scale. A summery of the ferrous materials which are regularly treated was given by /12/. The mechanical properties and corrosion resistance of the materials can be considerably improved by plasma nitriding. It was shown that while the nitrogen rich compound layer is responsible for the improvement of the wear and corrosion resistance, the diffusion layer influences mostly the fatigue properties.

Plasma nitrocarburizing is another type of plasma diffusion treatment which is very beneficial for various applications. The aim of the treatment is generally to enhance the wear and corrosion resistance of plain carbon steels, low alloy steels and cast irons through the production of an ε-carbonitride layer on the surface. Rie et al. /13/ have shown that the resistance to corrosion fatigue can be remarkably increased by plasma nitrocarburizing of low alloyed steels (fig. 1 - 3).

At present the most nitrocarburizing is carried out industrially using molten salt baths which allow an easy access to a mass production. However, the limited variety of process parameters inherent to salt bath treatment restricts the use of the process for production of defined microstructures. Plasma nitrocraburizing is indeed an alternative method to the conventional salt bath treatment.

Plasma nitriding of high alloyed steels has been investigated by /14/. Although stainless steels are widely used in chemical and vacuum industry the poor surface hardness and wear resistance often restrict their use in many cases. Plasma nitriding forms CrN and improves considerably the hardnes, however a reduced corrosion resistance is often associated with such treatment. Dearnly et al. /15/ have shown that proper treatment at low temperature can increase the hardness imparting excellent corrosion resistance. Therefore an increased use of plasma nitriding of stainless steels is expected.

The validity of the plasma carburizing was demonstrated in the past by many workers /16/. It is claimed that the enhanced mass transfer and the oxide free carburizing are the most significant advantages over other conventional carburizing processes (fig. 4).

Recently several full scale plasma carburizing units have been installed. The market potential is certainly larger than that of plasma nitriding. However, the engineering problems associated with integrating a quench unit with the plasma chamber should be solved satisfactorily.

Plasma boriding has been investigated to increase the surface hardness and to improve the resistance to high temperature oxidation /17/. By means of plasma it was possible to reduce the treatment temperature to 600 °C. Fe_2B phase with characteristic morphology is responsible for high hardness and wear resistance. Despite the well recognized advantages inherent to plasma boriding treatment this technique has not acquired an industrial access, probably due to the toxic nature of the boron content gases. Since the conventional boriding processes such as molten salt electrolyte or solid phase processes are hardly controllable, the plasma boriding technique is expected to replace the conventional boriding process in near future.

Nonferrous Materials

Although titanium alloys have many advantages of light weight and corrosion resistance, their engineering application is sometimes restricted because of the poor wear resistance and a high coefficient of friction. Recently there has been very intensive research to improve the tribological behaviour of titanium alloys by plasma nitriding /18/. The results obtained clearly show that plasma nitriding of titanium alloys has unequivocal advantages over gas nitriding:

1. The process rate can be increased.

2. The treatment temperature can be lowered.

3. ε-Ti_2N phase can be produced additionally to the high hardness TiN phase. ε-Ti_2N phase is known to be an effective barrierto the diffusion of ionized hydrogen.

By variation of process parameters it is possible to obtain different proportion of TiN and Ti_2N in compound layer. In aluminium content titanium alloys Ti_2AlN phase can be formed beneath the TiN and Ti_2N layers. Ti_2AlN phase is characterized by both high hardness and high ductility (fig. 5 and 6).

It is reported that automotive engineering components in Ti6Al4V have been sucessfully used after plasma nitriding. Rie et al. /19/ have developed plasma nitriding technique for titanium alloys to improve the wear and corrosion resistance as well as the biocompatibility of the knee-endoprosthesis /20/ which have been found to perform satisfactorily in service (fig. 7).

Plasma nitriding of Al and Al alloys is a very promising treatment in the automotive sector. Several investigations have been performed in the past /21/, and the difficulties associated with Al_2O_3 layer seem to be solved. Uniform AlN compound layer was successfully produced.

Chemical industries are especially interested to improve the resistance to corrosion and corrosion fatigue of nonferrous materials such as Zr, Hf, V, Nb, Ta, Co, Mo, TZM and W. Only few investigations on plasma nitriding have been carried out until now /22/, although the potential for future application is well recognized.

3. Lack of Knowledge-Suggestions for Future Research

There exists still considerable lack of knowledge in both basic mechanism and process technology.

Plasma Diagnostics

Unfortunately the plasma diagnostic investigations in the past were mainly concerned with gas mixtures which were not relevant to plasma nitriding or plasma nitrocarburizing.

Characterization of species and their population in plasma have to be investigated as a function of gas composition, gas pressure and plasma parameters. This kind of investigation will help us to understand better the mass transfer mechanism and will allow us to increase the layer growth rate without having detrimental structure deficiency. In view of the advantages inherent to the pulsed DC plasma process it is urgently needed to carry out plasma diagnostic research using pulsed DC plasma.

The effect of gas mixture and the ratio of pulse duration/pulse repitation time on nitriding efficiency has to be studied. Fig. 8 shows the preliminary resulsts obtained, recently using pulsed DC /23/.

The mechanism of layer formation during plasma nitriding is still matter of dispute, while the internal diffusion of interstials as a rate controlling process seems well established.

In case of PM materials, the layer formation and internal diffusion mechanism are more complicated than in compact materials. Therefore further study on the effect of pores and boundaries on plasma nitriding will be necessary (fig. 9).

In-situ Analysis

For better reproducibility of surface layers it will be a great help to carry out an in-situ analysis of surfaces during plasma nitriding. A versatile in-situ analysis method has to be developed. Plasma environment is certainly a disturbing factor which should be overcome. The combination of plasma diagnostic studies and in-situ analysis will allow us to optimize the process and to avoid unintentional incorrect treatment.

Microstructure and Properties

Effect of alloying elements on layer formation and growth rate has to be investigated more extensively. This will lead to a proper material selection for plasma nitriding.

Internal stress in connection with plamsa nitriding is not fully understood. While the wear resistance is mainly influenced by compound layer with high hardness, the internal stress resulting from the diffusion layer is responsible for improvement of fatigue properties. For rolling contact the internal stress is a decisive factor for life time increase. Relationship between materials composition, gas mixture and process parameters to optimize equally the surface hardness and internal stress has to be extensively investigated.

In many cases high wear and corrosion resistance alone is not sufficient for machine components. Friction coefficient plays often very important role and can be varied by plasma diffusion treatment. Amorphous carbon layer can be easily formed on the compound layer by manipulating the plasma parameters.

It is often claimed that plasma nitrocarburizing is an alternative to increase corrosion resistance of ferrous materials. Electrochemical measurement of the effect of monophases and their mixtures γ', ε and $(\gamma'+\varepsilon)$ on corrosion resistance is urgently required. It is expected that the cost extensive coating process for corrosion protection can be replaced partly by plasma diffusion treatment in case of ferrous materials depending on the requirements.

Little is known about the adherence of the hard coating produced by PVD, CVD or electrochemical deposition on the plasma nitrided layer. To develop the so called duplex-process or combined process it is necessary to understand better the cohesion behaviour between layers.

In the future increased application of multi layer system is expected. Combination of different surface treatment and coating processes is envisaged by the industries. It is emphasized that fundamental knowledge on microstructure and properties of such composite layers are prerequisit for the proper development of the adequate process combination /24/. Examples are given in fig. 10 and 11.

Process Development-Process Control

One of the most impressive development since last 10 years might be the microprocessor based control and micro-computer based process automation system. Thus the reliability has increased tremendously and the reproducibility is visibly improved. Nevertheless, there are still problems to be solved. Uniform temperature distribution is one of the crucial problems. Proper treatment of the geometrically different forms, sometimes with narrow holes should be thoroughly worked out.

Contineous processing is another challenging task for equipment manufacturers. In that way cost reducing process will be provided in the future for mass production.

It should be added that an extensive study on pre- and post-treatment of plasma diffusion treatment is required for high quality surface layers.

4. Conclusions

1. Extensive plasma diagnostic approach for mass transfer and layer formation mechanism is urgently needed.
2. In-situ analysis is required for better reproducibility, for which sensors have to be developed.
3. Unique relationship between process parameters and microstructures/properties of the layer has to be established. Consequently more extensive fundamental research is needed.

4. Increased effort for use of the micro-processor and micro-computer based process, control and automation system is required for easy access to routine work.
5. Multi-layer systems combined with other surface coating techniques will be increasingly requested in wide varieties of industries.
6. Future potentials and emerging fields of plasma diffusion treatment are discussed showing the trends in new applications.

References

/1/ K.-T. Rie (ed.): Plasma- und ionenstrahlunterstützte Verfahren für die Oberflächen- und Dünnschichttechnologie; Studie zur Analyse und Bewertung des derzeitigen technischen Standes und Strukturierung eines zugehörigen Forschungsprogrammes, Bundesministerium für Forschung und Technologie (BMFT), 1987 – 1988

/2/ J. Szekely, D. Apelian (eds.): Plasma Processing and Synthesis of Materials; Proc. of Symposium held in Boston, USA, 1983, Elsevier Science Publ. N. Y.

/3/ G. G. Tibbet: Role of Nitrogen Atoms in Ion-Nitriding; J. appl. Phys. 45 (1974), 5072 – 5073

/4/ A. Szabo, H. Wilhelmi: Zum Mechanismus der Nitrierung von Stahloberflächen in Gleichspannungsglimmentladungen, Härterei-Tech. Mitt. 39 (1984), 148 – 151

/5/ Yu. M. Lakhtin, Yu. M. Krymskii: Physical Processes in Ionic Nitriding; Protective Coating on Metals 2 (1970), 179 – 181

/6/ M. Hudis: Study of Ion Nitriding; J. appl. Phys. 44 (1973), 1489 – 1496

/7/ C. K. Johnes, S. W. Martin, D. J. Sturges, M. Hudis: Ion Nitriding; Proc. of the Conf. Heat Treatment 73, Metals Society London (1975), 71 – 75

/8/ J. Kölbel: Die Nitridschichtbildung bei der Glimmnitrierung; Westd. Verlag Köln (1965), Forschungsberichte des Landes Nordrhein-Westfalen, Nr. 1555

/9/ K. Keller: Schichtaufbau glimmnitrierter Eisenwerkstoffe; Härterei-Tech. Mitt. 26 (1971), 120 – 130

/10/ A. Ricard, J. Oseguera, H. Michel, M. Gantois: Exited State of Plasma for Steel Surface Nitriding; Proc. 1st Int. Conf. on Plasma Surface Engineering, Garmisch-Partenkirchen, FRG, 1988, DGM (1989), 83 - 90

/11/ A. C. Dexter, T. Farrell, M. I. Lees, B. J. Taylor: The Physical and Chemical Processes of Vacuum and Glow Discharge Carburizing; Proc. Intern. Seminar on Plasma Heat Treatment, Senlis, France, 1987, PYC Edition, Paris (1987), 53 - 71

/12/ W. Rembges: Fundamentals, Applications and Economical Considerations of Plasma Nitriding; Proc. Intern. Conf. on Ion Nitriding, Cleveland, 1986, ASM International (1986), 189 - 198

/13/ K.-T. Rie, Th. Lampe: Schwingungsrißkorrosion an plasmanitriertem Stahl; Werkstoff und Korrosion 33 (1982), 647 - 653

/14/ K. Ichii, K. Fujimura, T. Takase: Structure of the Ion-Nitrided Layer of 18-8 Stainless Steel; Techn. Report of Kansai University, Japan, 27 (1986), 135 - 144

/15/ P. A. Dearnley, A. Namvar, G. G. A. Hibberd, T. Bell: Some Observations on Plasma Nitriding Austenitic Stainless Steels; Proc. 1st Int. Conf. on Plasma Surface Engineering, Garmisch-Partenkirchen, FRG, 1988, DGM (1989), 219 - 226

/16/ K.-T. Rie, Th. Lampe, St. Eisenberg: Plasma Carburizing and Plasma Carbonitriding - An Austenitic Thermochemical Surface Treatment; Proc. 1st Intern. Conf. on Surface Engineering, Brighton 1985, The Welding Institute, Paper 39 (1985), 121 - 132

/17/ P. A. Dearnly, T. Farrell, T. Bell: Developments in Plasma Boronising; ASM J. for Energy Systems 8 (1986), 128 - 131

/18/ E. Rolinski: Isothermal and Cyclic Plasma Nitriding of Titanium Alloys; Surface Engineering 2 (1986), 35 - 42

/19/ K.-T. Rie, St. Eisenberg: Mikrostruktur von plasmanitriertem Titan und Titanlegierungen; Härterei-Tech. Mitt. 42 (1986), 344 - 348

/20/ K.-T. Rie, F. Schnatbaum: Optimieren der Parameter bei der Randschichthärtung von Bauteilen mit Plasmadiffusionsbehandlung - Möglichkeiten und Grenzen; to be published in "Maschinenmarkt"

/21/ T. Arai, H. Fujita, H. Tashikawa: Ion Nitriding of Aluminium and Aluminium Alloys; Proc. Intern. Conf. on Ion Nitriding, Cleveland 1986, ASM International (1986), 37 - 41

/22/ F. Matsuda, K. Nakata, K. Tohmoto: Ion Nitriding of Non Ferrous Alloys (Report I); Transactions of JWRI, Vol. 12, No. 2 (1983), 111 - 116, Welding Research Institute of Osaka University

/23/ K.-T. Rie, J. Wöhle: Spectroscopic Studies on Pulsed DC Plasma Nitriding of Steels; to be published in "Surface Engineering"

/24/ K.-T. Rie, Th. Lampe, St. Eisenberg: Abscheidung von Titannitridschichten mittels Plasma-CVD; Härterei-Techn. Mitt. 42 (1987), 153 - 159

Plasma Thermal Process	Plasma Diffusion Treatment (Thermochemical Process)	Plasma Assisted Deposition	Ion Beam Technique	Diverse Plasma Assisted Processes

– Sintering
 	
– Spraying
	– Carburizing
	
	– Boriding
	
	– Nitriding
		– PVD
		  – Ion Plating
		  – Sputtering
		– Plasma–CVD
		
		– Plasmapoly-
		  merization
			– Ion Implantation
			
			– Ion Beam Mixing
			
			– Ion Beam Assisted Deposition
				– e. g. Plasma–Etching

Table 1: Fields of Plasma and
Ion Surface Engineering

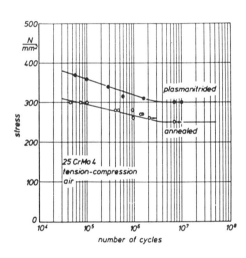

Fig. 1: Fatigue Properties of 25 CrMo 4
after Plasma Nitriding/Tension-
Compression

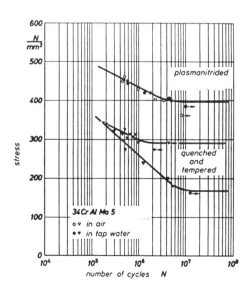

Fig. 2: Fatigue Properties of 34 CrAlMo 5
in Air and Tap Water after
Plasma Nitriding/Cyclic Torsion

51

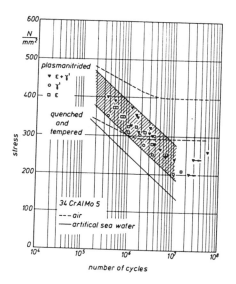

Fig. 3: Fatigue Properties of 34 CrAlMo 5 in Air and Artificial Sea Water after Plasma Nitriding/Cyclic Torsion

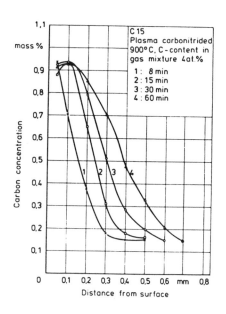

Fig. 4: Carbon Profiles of C 15 after Plasma Carburizing

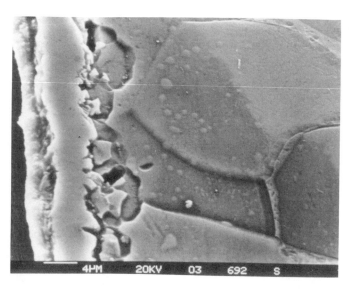

Fig. 5: Microstructure of Ti6Al4V after Plasma Nitriding

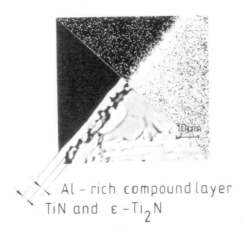

Al - rich compound layer
TiN and ε -Ti₂N

Fig. 6: SEM and Microprobe Analysis of
Ti6Al4V after Plasma Nitriding

Fig. 7: Knee-Endoprosthesis: Untreated
and Plasma Nitrided

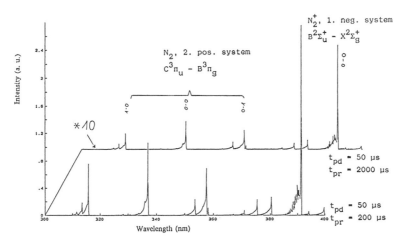

Fig. 8: Emission intensity vs. wavelength,
variation of pulseduration-/pulse-
repetitiontime, U = 550 V, 80 % N_2 / 20 % H_2

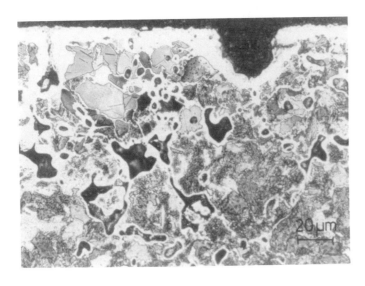

Fig. 9: Microstructure of Prealloyed
Sintered Steel SUMIRON¹ after
Plasma Nitrocarburizing

Fig. 10: Plasma CVD of TiN after
Plasma Nitriding of S 6-5-2

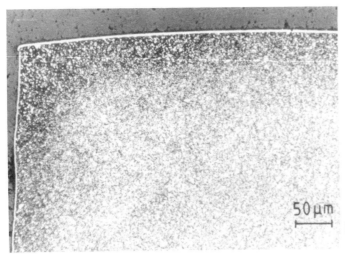

Fig. 11: PVD of TiAlN after
Plasma Nitriding of S 6-5-2

COMBINED PLASMA NITRIDING WITH PVD-HARD-COATINGS ON STEEL AND TITANIUM SUBSTRATES

Reinar Grün, Detlev Repenning
Plasma Technik Grün GmbH
Siegen, FRG

pvd-hardcoatings, e.g. TiN, TiC, CrN (Ti,Hf)N and other, are deposited on metallic substrates in layer thickness of 1 to 20 microns. Under high Hertzian pressure or similar loadings the pvd-layers fail due to the fact of high difference in hardness of the substrate material and the hardcoating. Substrate hardness normally is in the range of 300 to 850 HV. As a new possibility to withstand high loading wear the combination of puls plasma nitriding and pvd-hardcoating will be discussed in the paper.

The ion nitriding with adc-glow discharge by using short current pulses has some advantages in preparing a steel or titanium substrate for the following pvd-hardcoating. Those the producing of a clean surface without any nitride interlayer the low treatment temperature of the workpiece and last not least the homogenous character of the treated surface.

Specially on cold and warm working steels good results have been achieved in different applications as for components of pumps for the offshore technique, as for tools, for knives and others.

Different hard coatings are applied on the surface, which are summerized in table I. Results are shown in the diagram 1.

| material | plasma nitriding | | PVD coating | | | scratch test (N) | workpiece |
	depth (µm)	hardness (HV 1)	coating	thickness (µm)	hardness (HV 0.01)		
HSS	–	–	TiN	2-8	2 500	60 ± 10	
HSS	50	1400	TiZrC	8	2 900	90±10	cutter for chip panels
HSS	30	1100	CrN	2	2 000	80	cutter blade for plastics
HSS	30	950	TiWN	3	2 600	90	cutter blade for rubber
1.4462	240	1200	CrN / TiZrC	25	2 300	–	pumps for offshore
1.4493	100	1100	CrN CrN/i-C	20	3 500	–	gliding rings

Table I

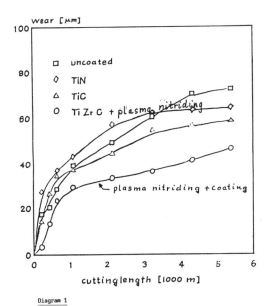

Diagram 1

Futhermore experiences have been gathered with gliding rings, which are used under high Hertzian pressure.

The application in table I is meanwhile industrial praxis. Especially the use of the combined processes for cutter brings considerable unprovements, which is illustrated in picture I.

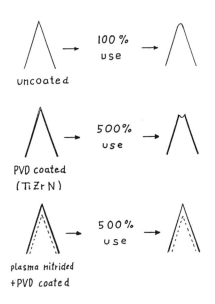

uncoated

PVD coated
(Ti Zr N)

plasma nitrided
+PVD coated

Picture I

56

A REVIEW OF PLASMA SURFACE MODIFICATION: PROCESS, PARAMETERS, AND MICROSTRUCTURAL CHARACTERIZATION

O. T. Inal, K. Ozbaysal, E. S. Metin, N. Y. Pehlinvanturk
Materials & Metallurgical Eng. Dept.
New Mexico Institute of Mining & Technology
Socorro, New Mexico, 87801, USA

Abstract

The main difference between glow-discharge nitriding and carburizing and the conventional practices is the instantaneous compound formation achieved in the plasma-assisted processes. Conventional processing is slowed down by a nucleation and growth process in compound-layer formation. In plasma processing, the compound layer is seen to form from sputtered surface atoms redeposited onto the surface with plasma gas atoms. Metastable phases rich in nitrogen are seen to form readily during ion nitriding of iron-based systems, thus affording more nitrogen available for diffusion. Even in systems where the activation energy for diffusion is not different in plasma versus conventional processing, the quickly attained surface saturation is seen to result in faster case growth kinetics.

Case growth rate in most systems studied is seen to be controlled by volume diffusion. Case depth is seen to be an inverse function of the concentration of the alloying species present and their affinity for the enriching species deposited from the plasma. The case/core boundary is also seen to be influenced by this affinity, with higher attraction resulting in sharply defined microfeatures and microhardness profiles.

The manuscript defines salient features of plasma processing of various engineering materials. Examples are given of experiments and characterization techniques that summarize our previous work in these areas.

LOW-ENERGY-GLOW DISCHARGE processing of metallic surfaces may result in the formation of metastable phases comprised of the plasma gas and surface species. These metastable phases play an important role in the control of diffusion processes required for the subsequent precipitate hardening of the host surface. The ease of dissociation of the metastable phase and the richer stoichiometry (in the metastable phase) of the plasma species for diffusion into the treatment region render plasma treatment much enhanced, in terms of an increase in reaction rate, over conventional surface modification procedures. The presence of a sputtering-generated point-defect population and its contribution to the increased reaction rate, are, however, still debated. The damage depth is so limited (of the order of a few tens of angstroms) that the temperature associated with the nitriding process is, in many cases, high enough to eliminate any significant contribution from the defects. The surface sputtering that takes place during glow-discharge processing does, however, allow for the formation of metastable phases rich in nitrogen or carbon and the immediate saturation of the surface; this is the most significant factor in terms of enhancing the kinetics of the process.

Electric-discharge plasmas utilized in ion nitriding and ion carburizing practice are created by the inelastic collision of the free electrons produced by the externally applied field (500–1000V), with the electron cloud of the gas atoms (molecules) utilized in the chamber at low (1 to 20 Torr) pressures. Since these free electrons interact almost entirely with the electron cloud of the heavy particle, and thus the ingredient that controls the molecule's chemistry, abundant active species are generated that can stimulate numerous chemical reactions at the host surface. Among the reactions assumed to take place with plasma exposure of surfaces, the following should be assessed for their significance in plasma-assisted nitriding and carburizing:[1]

1, sputter desorption of adsorbed impurities, which affords cleaner substrate surfaces for chemical reactivity;

2, ion impact enhances dissociation of adsorbed reactant species (a required process previous to adsorbed atoms being active), for enhanced chemical reactions that lead to surface compound formation;

3, generation of point defects that promote internal diffusion within the surface layers, though this effect is very limited in the case of ion nitriding/carburising due to the low energy of the ions;

4, enhanced incorporation probabilities for the bombarding ions, though again not a factor in the low-energy regimes considered here;

5, recoil implantation of surface atoms which reduces to a redeposition onto the substrate in the case of low-energy plasmas.

These features will be highlighted in this manuscript, in terms of various efforts we have made to date utilizing glow discharge plasmas. Studies on ion nitriding of low alloy steels, tool steels, maraging steels, stainless steels, PH stainless steels, titanium alloys and ion carburizing of steels and tungsten will be reviewed in terms of processing parameters, mechanisms envisioned for the formation of the associated metastable hardening phases, and precipitate nucleation and formation. Characterization techniques utilized will be summarized and salient features in each case will be indicated.

FEATURES OF PLASMA TREATMENT

PROCESS PARAMETERS - Ion nitriding - Low-energy glow-discharge exposure of metallic surfaces is practiced in ion nitriding of alloy steels to improve wear resistance and fatigue strength as indicated earlier. The increased hardness of the surface layers is attributed to the formation of a fine dispersion of coherent or semi-coherent nitrides of the alloying elements present in the steels. The rate of this reaction is believed to depend on the strength of interaction of the particular alloying element and nitrogen, the ease with which the precipitates nucleate and grow, the concentration of the alloying elements, the nitriding potential of the gas mixture used, and the temperature of nitriding[2]. The nitride-forming alloy species (Ti, V, Cr, Mo, etc.) are also carbide formers and the corresponding carbides have sufficient time to form in the pre-nitriding tempering step. Thus the amount of carbon present in the steel should affect the rate of nitriding through an additional stage that involves carbide dissolution prior to nitride formation[3,4]. Another influence of carbon content would, of course, involve the type of microstructure attained following the quench and tempering, and the way that influences the nitrogen diffusion during ion nitriding. Our results on the influence of each parameter will be detailed below.

Two types of hardness vs depth curves are generally observed in ion nitriding of steels. The first type is characterized by the formation of a uniformly hard subscale which advances progressively into the core region with time, and is exhibited by steels containing strong nitride formers (such as Ti, Al and V). The second type is characterized by a comparatively smooth transition from maximum hardness near the surface to core hardness, at a depth dependent on both time and temperature, and is exhibited by steels with weaker interaction between alloying elements (such as Cr and Mo) and nitrogen. In the latter case, the rate of hardening is controlled by the rate of reaction to form the nitride phase as well as by the bulk diffusion of nitrogen through the sample. The result is that the case-core interface is less abrupt[3,4]. In addition, the presence of two or more nitridable species is seen to afford much higher peak hardness values than equal amounts of either species alone; this point will be elaborated later[5].

In ion nitriding practice of steels, the standard pre-nitriding procedure involves solution treatment, quench and tempering steps to render the sample amenable to easy nitrogen diffusion. As stated above, the nitridable alloying species (except for Al) are also strong carbide formers, and the carbides have sufficient time to form in the tempering step; it is normal therefore to expect a carbide dissolution reaction previous to the nucleation and growth of the nitride precipitates. This has indeed been experimentally verified by our field-ion microscopy study that conclusively shows the nitride precipitates to have formed in the vicinity of undissolved carbide clusters[3], replacing carbide particles that were observed following tempering. Phillips and Seybolt[6] also observed that the coarse carbides, formed during the tempering of their steels, had partially dissolved during ion-nitriding in order to allow for the nucleation and growth of the nitrides; they assumed, however, that this has little bearing on the nucleation and growth rate of nitrides. This latter assumption is in contrast with our findings which conclusively show that it does matter whether the nitridable species is attached or is free to accept the nitrogen[4]; the same peak hardness was achieved in steels following nitriding for long duration (48h) and at high temperature (520°C), although one contained much less of the nitridable alloy species than the other two but also had much less carbon.

Microhardness and metallographic examinations show that microhardness of the case increases with increased content of the nitridable alloy species while case depth decreases, for a given time and temperature of treatment[3-5]. This is true for both single and combined additions of nitridable species, though the microhardness values achieved are greater for the combined additions, as indicated earlier[5].

It is a generally accepted fact that a mixture of hydrogen and nitrogen gases, with nitrogen proportions as low as two volume percent, to eliminate undesirable compound-layer formation, is ideal for ion-nitriding purposes. Our studies on pure nitrogen versus a nitrogen-hydrogen mixture (20 to 80 volume percent, respectively) for ion nitriding of pure iron indicate the production of a thicker compound layer in pure nitrogen but thicker case depths in the mixture, which supports this thinking[7]. The catalytic role played by hydrogen in terms of the formation of compounds richer in nitrogen (Fe_2N) on the surface is possibly responsible for the thicker case depths. Pure nitrogen, on the other hand, seems to afford higher kinetics of compound layer growth because of the higher partial pressure of nitrogen present. This point will be elaborated upon in the next section.

Since ion nitriding is a diffusion-related process, increases in treatment temperature lead to increased compound-layer thickness, and increased case depths for a given treatment time, but an overall decrease in the peak hardness values achieved[3,4]. As with age-hardening alloys where a maximum in hardness is achieved at a particular aging temperature and time the maximum hardness achieved by ion nitriding is based on optimum precipitate density and size; these can be lost at higher temperatures and longer times due to precipitate growth and further tempering of the matrix.

Other-plasma related parameters such as power and current density seem to play very negligible roles in ion nitriding. High power input only enhances the chances of arcing and should be avoided. Since the rate-limiting step in ion nitriding is the introduction of nitrogen into the steel, either the rate of nitrogen diffusion through the matrix or the rate of nitrogen transfer at the steel/compound-layer interface should control the kinetics of this process; compound-layer formation occurs readily in ion nitriding thus fixing the nitrogen concentration at the surface of the diffusion layer at that of the Fe/compound-layer equilibrium. Thus, plasma current density should play no rate-controlling role in this process. This was verified in our experiments[4].

Ion-carburizing - As in the case of ion nitriding, the most important process parameters in ion-carburizing are treatment temperature and time. Our studies indicate that a surface carbon saturation occurs readily and is independent of the chemical composition of the substrate. Treatment temperature and time affect the rate and extent of carbon diffusion, and thus the associated features of case depth and hardnesses achieved. This also will be explained further in the next section.

METASTABLE PHASE FORMATION - Ion Nitriding - Fe-Based Alloys - Formation and growth of iron nitrides on pure iron in nitrogen and nitrogen/hydrogen (20%/80%) plasmas have been studied at temperatures between 500-600°C. The following differences between the results for the two plasma compositions, and comparisons with the results obtained with conventional processing[7] are summarized below.

a. Ion-nitriding with the N_2/H_2 plasma, involved the ready formation of (Fe_2N) phase at the surface, which decomposed into (Fe_4N) or ($Fe_{2-3}N$) at longer treatment times, depending on the temperature of treatment. The latter two nitrides were the only species observed with pure nitrogen plasma, indicating that a catalytic action of hydrogen is a requirement for the formation of the nitrogen-rich Fe_2N phase. The small size of the initially formed nitrides would suggest a mechanism involving sputtering of iron atoms off the surface, compound formation in the cathode fall region, and redeposition onto the substrate surface as verified by:

i. the small particle size of the nitrides as indicated by SEM study, and

ii. the saturation of the nitride-layer formation rate with increased time. (This follows a highly accelerated initial growth, possibly controlled by the high initial rate of Fe sputtering from the clean surface versus an increased mean free path of iron atoms to the cathode fall region with compound coverage of the substrate surface). In the conventional process[8] the growth of the compound layer is parabolic and much slower initially, but the compound-layer thickness after longer times is similar in the two processes[7].

b. Immediate formation of the nitride layers on the substrate surface, resulting in the build-up of a large nitrogen concentration gradient and their vapor-deposited appearance, is thought to be the basic reason behind the enhanced rates achieved with ion nitriding over the conventional use of a gaseous medium[3,4]. Free nitrogen supplied by the readily occurring decomposition of ξ phase may also be a contributing factor in this regard, as also pointed out by Edenhofer[9].

c. Nitride layer thicknesses obtained with the pure nitrogen plasma were seen to be slightly lower than with the N_2/H_2 plasma, possibly due to the presence of a higher nitrogen partial pressure in the former plasma, though in both cases this growth is not parabolic as in conventional nitriding[10] and occurs at a much higher rate initially.

d. Nitrogen/hydrogen plasma produced larger case depths than pure nitrogen plasma, possibly due to the initial presence of the Fe_2N phase in the former case and its rapid dissociation to contribute to enhanced nitrogen concentration. Case depth in both instances was parabolic with time, indicating a volume diffusion controlled process.

e. Activation energies calculated for case growth yielded lower values than the activation energy required for nitrogen diffusion in α iron. The difference is assumed to be due to non-parabolic growth of the nitride layer, and is therefore, accepted as a direct indication of the faster kinetics of nitriding induced by the plasma treatment.

Titanium and Titanium Alloys – Interstitial elements such as oxygen, carbon, and nitrogen are very strong α stabilizers and strengthen titanium alloys. Among these elements, nitrogen has the greatest strengthening effect[11] and thus nitriding of titanium has received more attention than the inclusion of any other interstitial species.

Titanium has a very high affinity for nitrogen and the formation of TiN is thermodynamically favored within the typical parameter ranges used in ion nitriding[12]. Equilibrium partial pressure, at the 800°C temperature of ion-nitriding practice, is 1.02×10^{-23} atm., which is actually much lower than the minimum nitrogen partial pressure used in typical ion-nitriding practice (1.3×10^{-3} atm.). Thus the formation of the stable TiN would dictate no difference in the kinetics of nitriding by gas or plasma techniques. This is shown to be true in our studies[12] and elsewhere[13], where plasma composition and current density at a given temperature are seen to have no effect on the nitriding kinetics. The advantage of ion nitriding over other nitriding practices, such as gas nitriding, for titanium and its alloys is therefore limited to the lower temperatures at which the formation of TiN at the surface is activated by the presence of plasma.

The initial TiN formation takes place earlier in the plasma, as indicated for the formation of iron nitrides. This also is supported by the SEM study of morphological features since the redeposited nitrides on the surface also exhibit features of a vapor-deposited film. Since the ion energies used are fairly low, the damage depth is quite low as well (typically less than 1 nm) and thus the structural alteration at the surface, in the form of increased point-defect concentration, is not sufficient to influence long-range diffusion process.

The unique feature of plasma nitriding of Ti and its alloys is the diffusion couple that forms between the TiN surface layer and the Ti core. With increase in the duration of ion nitriding, diffusional growth may result in the formation of several layers within this diffusion couple as dictated by the equilibrium phase diagram[14]. Depending on the processing temperature, as many as four phases may exist in this couple:

δ (TiN), ε (Ti_2N), α (Ti-N solid solution (hcp)), and β (Ti-N solid solution).

Titanium alloys may broadly be classified as α, $\alpha + \beta$, and β alloys, depending on the phases present. Regardless of the above distinction, a thin nitride exists on the surface upon exposure to nitrogen-bearing plasma. The thickness of the nitride layers (TiN and Ti_2N) is a function of processing temperature and time. In $\alpha + \beta$ alloys a nitrogen-stabilized layer appears under the surface nitrides. As the amount of nitrogen decreases towards the core, the amount of the predominant phases (α and β) approach their equilibrium contents.

Beta alloys, due to their more open bcc structure, allow for faster diffusion of nitrogen resulting in much larger case depths in these alloys. Considerable hardening has not, however, so far been obtained in β alloys. Since nitrogen stabilizes the α-phase, a region rich in α precipitates is observed in the nitrided case of these alloys. Although the precipitation of α contributes to the strength of β alloys, similar strength can be achieved by proper heat treatment, and thus ion nitriding does not contribute in any significant manner apart from the presence of the thin TiN layer at the surface.

The TiN surface layer contributes to high hardness (in excess of 1500 VHN). Nitrogen significantly increases the strength of the α phase at the expense of negligible ductility. Thus, large case depths of several hundred micrometers, as in the case of Fe-based alloys, should be avoided in Ti and Ti alloys. The major strengthening mode beyond this layer is the solid-solution strengthening experienced by the α phase in the presence of nitrogen. A thin layer containing Ti_2N precipitates also exists at the nitride/α interface of nitrided α and $\alpha + \beta$ alloys[13].

Ion implantation of Ti and Ti alloys with nitrogen has produced excellent wear properties due to a fine precipitate of TiN[15], and possibly due to the formation of a thin amorphous layer[16]. A comparative study, however, shows that gas nitriding affords better wear resistance with lower treatment temperatures (below 723°K) than ion implantation[15]. Since ion nitriding affords lower treatment temperatures yet than gas nitriding, it should compare more favorably with ion implantation[12].

ION CARBURIZING - Fe-Based Alloys - In principle, plasma carburizing or ion carburizing is analogous to ion nitriding. The specimen or workpiece is made cathodic with respect to the chamber in a d.c. electric circuit. Approximately 500-800 V will sustain the plasma once initiated in the carbonaceous atmosphere. It is within this plasma that the carburizing gas, e.g. methane (CH_4), is dissociated and thus a high carbon potential is established and maintained at the workpiece surface. The carburizing species bombard the cathodic workpiece and subsequently diffuse inwards to form a carbon-rich surface layer. In ferrous alloys, plasma carburizing is carried out in the austenitic phase region and produces a martensitic structure upon quenching. To avoid soot formation, it is a good practice to heat the workpiece to the process temperature in an H_2 environment with the glow discharge and introduce the carburizing gas at a 5% CH_4 - 95% H_2 proportion at a total dynamic pressure of 10 Torr for the processing[18].

Our studies[18,19] on ion carburizing of AISI 1020, 4620 and 8620 steels (AISI 1020 is a plain carbon steel whereas AISI 4620 and 8620 contain austenite-and ferrite-stabilizing elements respectively) reveal very little difference in the ion-carburizing behaviors exhibited, although compositional differences among them are extensive. A small difference in the activation energies for carbon diffusion in austenite is the only indication of this compositional difference. These calculated activation energies are also very close in value to those computed for gas carburizing[20]. Thus the increased case depths obtained in our studies cannot be attributed to increased diffusion rates but to a large carbon concentration gradient established in the very early stages of plasma carburizing. As in the case of gas nitriding, the carbon saturation of the surface in gas carburizing should have a nucleation and growth step that is time consuming. Our studies conclusively reveal that the spontaneous saturation of the surface with carbon occurs in the very early stages of plasma carburizing, regardless of the chemical composition of the steels[18,19]. The growth of case depth is parabolic with time for the steels investigated, suggesting a volume diffusion controlled process, as in the case of ion nitriding.

Tungsten - The solubility of carbon in tungsten at conventional carburizing temperatures is practically zero; thus no solution-hardening contribution was expected

in this study with carbon enrichment from plasma exposure. A compound layer, comprising tungsten carbides (WC and W_2C), is formed with the use of plasma, affording a very hard surface layer. The study was made in similar gaseous mixtures to those utilized for steels and the following observations were made[21]:

(i) a high carbon concentration is obtained, spontaneously, with exposure to plasma;

(ii) the process is seen to be interface controlled initially and volume-diffusion controlled in the later stages of carburizing;

(iii) highly uniform and crack free case regions are obtained with plasma carburizing.

Titanium - Ion carburizing of titanium and titanium alloys can be done in a plasma containing methane, with typical plasma-carburizing parameters[18]. The process yields free hydrogen upon dissociation of methane, which is prone to diffuse into the titanium-bearing matrix. Although a layer of TiC several microns thick can be obtained at the surface, the core is decorated with TiH_2 precipitates that embrittles the matrix[12]. Therefore, plasma treatment of titanium and its alloys in hydrogen-bearing media seems to be impractical.

EXPERIMENTAL OBSERVATIONS

This section will describe our specific applications of ion nitriding and carburizing, and the observations we have made on the plasma treatment of Fe-based, Ti-based, and W-based alloys.

ION-NITRIDING EXPERIMENTS - Low Alloy Steels - Our work on AISI 1040 steel, alloyed with 1,2 and 3% each of Cr, Al, and Cr + Al, was carried out to delineate the role of individual nitride formers compared to the combined presence of two nitridable species[5,22]. The prenitriding treatments given to the samples were similar except for the alloy containing 3% Al, which had to be cold rolled to 40% reduction to aid in nitrogen diffusion in the absence of a martensitic structure in the matrix. As expected, increased amounts of alloying species resulted in increased surface hardness for each alloy and the combined addition of the two species produced the highest hardness. The Cr containing steels exhibited lower hardnesses than the corresponding Al-containing steels with a smoother transition between the hardened case and the sample core, as indicated earlier for weaker nitride formers. The aluminum-containing steels exhibited a much sharper interface between the case and core regions, which was also seen in alloys containing Al plus Cr. This is consistent with our other work on ion-nitriding behavior. In nitralloy 135 M, that contains both Cr and Al, higher hardness, shallower case depths and well defined case/core interfaces were observed[3,4]. Atom-probe field ion microscopy showed the CrN to be uniformly distributed throughout the case region, while AlN was seen to be precipitated in defected regions, as expected from the precipitation of an hcp material in a bcc matrix. Large carbide particles were seen to exist in Cr-containing alloys even after 36 hours of ion nitriding, verifying the fact that the affinity of Cr for nitrogen is not strong enough to induce the dissolution of all carbide particles.

The higher hardness associated with the co-presence of two nitride formers can possibly be explained through the transmission-electron-microscope analysis made on these alloys. AlN precipitates are seen to be larger when CrN nitride is also present, in the case regions, suggesting that AlN nucleates earlier and grows to a larger size in the presence of Cr. It seems likely that Cr has some sort of a reducing effect on the activation energy for AlN precipitation. This may come about if the activity of Al increases (which would make the free energy of AlN formation more negative), or if the presence of Cr brings about a decrease in the strain energy for AlN precipitation. The atom-probe data show a lack of uniformity of CrN distribution in the case region, that was not the case with alloys of Cr only, and the CrN precipitates are seen to coexist in the same defected regions as the AlN precipitates thus giving support to the latter assumption.

Tool Steels - Tool steels contain temperature-stable carbides of elements such as tungsten, vanadium, and chromium which render them quite hard and wear resistant. These steels were studied since ion nitriding was thought to improve on the wear properties without the introduction of extensive distortion and core softening; the studies were designed to elucidate the role played by carbon in the ion-nitriding of alloyed steels. This study utilized H13, M2, D2, and D3 steels with process variables of temperature, time, and partial pressure of nitrogen in the plasma gas. Also to be scrutinized was the role of the carbide-dissolution/nitride-formation sequence in these alloys containing species of low (W and Mo), medium (Cr), and high (V) affinity for nitrogen.

The highest nitrided hardness was observed in the M2 and H13 steels. This is assumed to be due to their lower carbon content (1.00 and 0.35 wt.%, respectively), which influences the efficiency of nitriding through: i) occupying the nitridable alloy species through carbide formation in the tempering stage, and ii) alteration of nitrogen diffusivity within the matrix. These steels also contained V, which is a strong nitride former (2.00 and 1.00 wt%, respectively), which accounts for the higher surface hardness. D2 and D3 both exhibited lower surface hardness and smaller case depths. These steels contain higher carbon content (1.50 and 2.25 wt%, respectively) and alloying species with moderate to low affinity for nitrogen (D2: 1.00%Mo, 12%Cr; D3:12%Cr). The large carbide particles observed in these steels following tempering obviously occupied the alloying species and reduced nitriding efficiency with the result of relatively lower surface hardness and shallower case depths. All four steels showed increasing case depth and decreasing surface hardness with increasing process temperature and time. Both properties were also found to be independent of nitrogen partial pressure in the plasma for all steels except the M2. The lower carbon content of this steel, together with the presence of large quantities of nitride forming species (5%Mo, 4%Cr, and 2%V) along with a strong carbide former (6%W), left a sufficient amount of the nitridable species unattached which in turn increased nitrogen diffusivity; thus the consumption rate for nitrogen was high. Although, in general, nitrogen saturation of the surface is readily achieved in ion nitriding, as indicated earlier, it required higher nitrogen partial pressures to increase case depths in this steel. The higher efficiency of nitriding is evidenced by the larger case depths and higher hardnesses achieved.

The activation energies for nitriding were calculated for H13 and D3 steels as 19.5 Kcal mole^{-1} and 28.5 Kcal mole^{-1} respectively, elucidating the slower reaction rate associated with the high carbon steel (D3) and the definite role played by carbon in nitrogen diffusivity and nitride formation. Thus, nitrogen diffusivity is composition dependent and the nitriding reaction is slower in high carbon steels.

X-ray-diffraction analysis showed extensive line broadening of major martensitic peaks, and Fe-Cr carbides were seen to convert to chromium nitrides whereas Fe-W carbides were stable in the nitrided case. Wavelength-dispersive spectroscopy revealed that Fe-Cr carbides were converted to chromium nitrides in regions close to the surface, whereas conversion was not complete at larger depths (greater than 80 μm)[23].

Stainless Steels - Because of the high Cr content they possess, all stainless steels can be ion nitrided to some degree. These steels, that are used in applications requiring corrosion resistance, are assumed to be difficult to nitride because of the surface oxide film. This film has to be removed through a pickling operation so that nitrogen diffusion is not hindered. Although corrosion resistance is improved in most steels following plasma surface modification, the effect on stainless steel is expected to be adverse[24]; nevertheless the increase in surface hardness, the lower coefficient of friction and the improved abrasion and fatigue resistance makes this practice worthwhile.

This study was made to characterize the structure sensitivity of nitrogen diffusion and the corresponding dependence of nitriding behavior for martensitic (410), ferritic (430), and austenitic (321) stainless steels[28] on the process variables of temperature and time. Two different types of nitriding behavior were observed. The first, exhibited by the martensitic (bct) and ferritic (bcc) stainless steels, is characterized by faster formation of the diffusion layer that advances progressively into the core with time. The second type, exhibited by the austenitic stainless steel (fcc), shows a comparatively slower formation of the diffusion layer. The difference in the observed diffusion behavior of nitrogen arises from the structure dependency of nitrogen diffusion[26]. The activation energy for nitrogen diffusion is known to be higher in fcc austenite (47.3 kcal/mole) than it is in bcc ferrite (18.3 Kcal/mole)[8,27]. A slight superiority observed with martensitic grades over ferritic grades can be explained with the aid of the structure sensitivity of interstitial diffusion[26]. Since the ion-nitriding temperatures utilized were relatively low, crystal defects are expected to play an important role in the diffusion of nitrogen. The small grain size and high dislocation density that exist in the martensitic structure probably acted as short diffusion circuits for nitrogen diffusion. This leads to the larger case depths observed. The presence of titanium in the austenitic stainless steel may also have contributed to the lower case depths observed in this material. Titanium displays a stronger affinity for nitrogen than Cr[3]; thus it is reasonable to assume that, since the carbon content is low (0.08 wt %; Ti content is 5 times that of C) in this steel, faster nitrogen consumption occurs readily. This is verified by the fact that, at 480°C, the austenitic steel develops a higher hardness than the others, and it also suffers a smaller decrease in surface hardness with increasing ion-nitriding temperature than the other two steels; both the TiN and TiC formed are stable compounds.

No structural alteration is observed in the core region of ferritic and austenitic stainless steels following ion nitriding, whereas evidence of tempering of the core occurred in the martensitic stainless steels. Our work on PH stainless steels, incomplete and unpublished as yet, seems to follow very similar trends, where austenitic steels exhibit slower rates of ion-nitriding behavior, with a slight increase in semi-austenitic steels and much higher rates observed in martensitic PH stainless steels.

Maraging Steels - Maraging steels are special-purpose tool steels that may be used for both cold and hot work applications. These steels are aged, after a solution heat treatment between 420 and 520°C; this procedure strengthens them by the aging of the bcc nickel martensite structure. The aims of this study were to elucidate the role of nitrogen enrichment in the case region in inducing a reversion to austenite during ion nitriding, and to characterize the increase in surface hardness induced by plasma treatment. For this purpose, maraging grade 250, 300, and 350 steels were ion nitrided with process parameters of time and temperature. All three steels were in the solution-treated and quenched condition prior to ion nitriding. Age hardening was allowed to occur during ion nitriding to avoid additional reversion of the structure to austenite and overall softening. The fact that nitrogen is a strong austenite stabilizer (30 times more effective than nickel) would only have increased this reversion process during the plasma-induced diffusion[22].

As expected, the highest hardnesses for both surface and core were achieved at lower nitriding temperatures (<480°C) and shorter ion-nitriding durations (<10 hours). Comparison of the slopes of curves of case depth versus plasma-treatment time, that would give reaction rate constants, indicate that reaction is fastest in grade 250 and slowest in grade 350; this is attributable to the latter's higher Ti content (0.30 for 250 and 1.42 for grade 350). (The role of Ti in ion-nitriding behavior was fully explained in the previous section). This was also verified by the higher surface hardness and shallower case depths observed in the latter steel.

Since all three steels contain large amounts of Ni (>18%) and also Al and Ti, the latter two being strong nitride formers, we expected them to contain Ni_3Ti and Ni_3Al; the ring patterns obtained showed[22] only the presence of Ni_3Ti, and our recent TEM work has indicated AlN to be present as well. The diffraction patterns of the core region are quite complex suggesting the presence of multiple phases, but the case region, after nitriding, exhibits clear rings of a bcc phase (possibly α), an fcc phase (likely to be γ) and TiN. In summary, benefit from ion-nitriding of maraging steels can be attained through careful avoidance of reversion to austenite[22].

Titanium and Titanium Alloys - Pure titanium and Ti 6242 alloy were ion nitrided between 800-1000°C for various durations (4-20 hrs) at 5 Torr total pressure under pure nitrogen-plasma. Layers of TiN and Ti_2N at the surface followed by a nitrogen stabilized α-Ti were observed to form. The growth of nitride layers was parabolic with time, again indicating a diffusion-controlled process[28].

Titanium undergoes an allotropic transformation between the low-temperature hcp (α) phase and the high-temperature bcc (β) phase, at 882°C. Therefore nitriding at temperatures above the β transus shows a nitrogen-stabilized α case and a transformed β successively, under the nitride layers. Ion nitriding at temperatures below the transformation temperature produces a nitride layer that can be seen as a gray band at the outer most edge and an equiaxed α - Ti core, with a white region at the nitride/α boundary and characteristic concentration dependent etching bands beneath[29].

Kinetics of nitride-layer growth on pure titanium were investigated at temperatures above and below beta transformation; similar experiments on Ti 6242 were made at temperatures below the beta transus due to the relatively high transformation temperature of this alloy (993°C). Ion nitriding of pure titanium resulted in a layered structure of titanium nitrides; Ti 6242 alloys, however, showed only the formation of the TiN layer with no Ti_2N layer underneath. Ti_2N was observed in Ti 6242 in the form of precipitates in the α phase, adjacent to the TiN/α interface. The thicknesses of the nitride layer were similar in both materials although the microstructural constituents were different.

Electron-microscopy observations of ion-nitrided pure titanium have shown that precipitation of Ti_2N takes place within the nitrogen-stablized α, with the morphology depending on the degree of initial supersaturation at the time of precipitation, as cooling takes place. In many instances, dislocations are formed due to diffusion-related stresses and are assumed to act as nucleation sites for these precipitates. Precipitation of Ti_2N was also observed in nitrided Ti 6242 alloy. Precipitation in this alloy was merely confined to α/β phase interfaces where crystallographically favorable sites exist. The morphology of precipitation, however, was different from that in pure titanium and appeared mostly striated.

ION-CARBURIZING EXPERIMENTS - Steels - Carburizing steels, AISI 1020, 4620, and 8620, were exposed to a 5 vol.% CH_4-95 vol.% H_2 plasma at a total pressure of 10 Torr in the temperature range 850-1025°C for various durations of ion carburizing. In order to reduce the carbon concentration at the surface to an acceptable level, plasma carburizing was followed by a diffusion anneal in a separate furnace, followed by oil quenching[18].

The growth of case depth was seen to be parabolic with time, as in the case of ion nitriding. The enhanced case-depth growth rates observed over gas nitriding[30] is believed to be due to the rapid saturation of the surface with carbon in plasma carburizing[18].

Tungsten - Sintered tungsten samples were ion carburized at 10 Torr total pressure in a 5% methane/95% hydrogen gas mixture in the temperature range of 900-1100°C for various durations of plasma exposure. A WC compound layer was obtained at the surface.

The initial linear increase in thickness of the compound layer formed with time indicates the process to be interface controlled in the early stages of carburizing. After an incubation period, the process is seen to be controlled by volume diffusion. A thermodynamically unstable phase, W_2C, was seen to be formed along with WC only at the lower carburizing temperature (900°C), suggesting the attainment of a high carbon concentration at the surface quite readily. This W_2C compound is seen to dissociate to WC at longer times or higher temperatures of treatment. Since no carbon diffusion into the substrate occurs, no sign of solution hardening was observed, as expected[21].

CHARACTERIZATION TECHNIQUES

To characterize the existence of the case and its depth, use of optical microscopy and microhardness depth profiling has been the common practice. Judicial etching following metallographic sample preparation has been used to advantage to differentiate the compound layer, case and core regions because of their different etching behavior. This information is usually supplemented by microhardness depth profiling with a minimum of 200 gf for these hardened cross-sections; the case depth is then defined as the region at which the hardness is 2 HRC above that of the core hardness. Since the hardness profiles also delineate the nature, amount, and distribution of the hardening precipitates they should be made in close promixity to one another for better definition.

Transmission electron microscopy has made very valuable contributions to the practice of plasma processing. We have learned the nature of precipitate/matrix habit correlations (coherent or semicoherent), and we have learned the need for the presence of dislocations and other defects for the precipitation of non-fcc (AlN-hcp) precipitates in the matrix following ion-nitriding. Selected area-diffraction information has elucidated the habit features of these precipitates and their distribution within the case region. Field-ion microscopy was the first technique that indicated the existence of a dual process in steels, that is of carbide dissolution versus nitride formation, following nitrogenous plasma exposure. This technique, together with atom-probe field-ion microscopy, has shown that CrN precipitates are uniformly distributed in the case region when alone but that they cluster to the same defected region when Al is also present; the latter situation provides the more efficient hardening. A good estimate of dimensional features of the nitrides formed has also been obtained by this technique.

Scanning electron microscopy, along with wavelength-dispersive spectroscopy, was very useful in characterization of the nitrides and carbides present in the case and core regions in our tool-steel study[26]. These techniques should also be very useful for qualitative studies of the distribution and concentration of nitrogen and carbon, especially in those instances where the advantage gained from plasma exposure is solid-solution strengthening based on enrichment of the interstitial species. X-ray diffraction was also used to advantage for these characterizations.

Finally, wear and friction testing are necessary to obtain application-oriented information. The choice of scriber has to be based on the microhardness achieved information obtained for the surfaces in question[4].

CONCLUSIONS

Use of plasma in nitriding and carburizing practice seems to have the advantage that surface saturation occurs more readily than in conventional processing. The formation of these saturated layers is seen to be based on the sputtering of substrate atoms, their joining with the enriching species in the cathode fall region to form a compound, and subsequent redeposition onto the substrate surface. This is verified by the scrutiny of their geometry with SEM, following plasma exposures of short duration. The nucleation and growth processes that are rate limiting in the initial phase of conventional processing do not occur with plasma exposure. In general, case growth is characterized by volume-diffusion control. Higher temperatures and longer durations of plasma exposure are seen to result in lower hardness but thicker case depths; the lower hardness is believed to result from precipitate coalescence. Greater affinity of the nitride-forming species for nitrogen leads to higher surface hardness and smaller case depths, with a sharp interfacial definition between the case and the core indicating a high efficiency of nitrogen consumption.

In summary, plasma-assisted nitriding and carburizing offer clean and efficient processing for improvement of wear and gall properties. They utilize lower temperatures and shorter processing times which in turn contribute to the better maintenance of core properties.

REFERENCES

1. J.A. Thorton, in Ion Nitriding, Proceedings of an International Conference on Ion-Nitriding, 15-17, September 1986, Cleveland, Ohio; T. Spalvins ed., ASM Int. p. 19.
2. P.C. Jindhal, J. Vac. Sci. Technol., 15 (2) (1978) 313.
3. O.T. Inal and C.V. Robino, Thin Solid Films, 95 (1982) 195.
4. C.V. Robino and O.T. Inal, Mat. Sci. and Eng., 59 (1983) 79.
5. O.T. Inal, K. Ozbaysal, N. Pehlivanturk and G.L. Kellogg, "Mechanism of Ion-Nitriding in Alloyed 1040 Steels; TEM, FIM and Atom-Probe Characterizations", submitted for publication (1989).
6. V.A. Phillips and A.U. Seybolt, Trans. Metall. Soc. AIME, 242 (1968) 2415.
7. E.S. Metin and O.T. Inal, J. of Mat. Sci., 22 (1987) 2783.
8. K. Schwerdtfeger, P. Grieveson and E.T. Turkdogan, Trans. AIME, 245 (1969) 2461.
9. B. Edenhofer, Heat Treat. Met., 1 (1974) 23.
10. K.H. Jack, Proc. Roy. Soc., A208 (1951) 216.
11. R.I. Jaffee, H.R. Ogden and D.J. Maykuth; Trans. AIME, 188 (1950) 1261.
12. E.S. Metin and O.T. Inal, to be published in Light Metal Age, (1989).
13. E.S. Metin and O.T. Inal, "Kinetics of Layer Growth and Multiphase Diffusion in Ion-Nitrided Titanium", to be published in Metallurgical Transactions (1989).
14. H.B. Bomberger, Technical Report, RMI Titanium Co., 1978.
15. R. Hutchings, Mat. Sci. and Eng. Lett., 1(1983)137
16. R.G. Vartdiman and R.A. Kaut, J. Appl. Phys., 52 (1982) 690.
17. R. Martinella, S. Giovanardi, G. Chevalland, M. Villani, A. Molinari and C. Tosella, Mat. Sci. & Eng., 69 (1985) 247.
18. N.Y. Pehlinvanturk, O.T. Inal and K. Ozbaysal, Surface and Coatings Technology, 35 (1988) 309.
19. N.Y. Pehlinvanturk, and O.T. Inal, Adv. Mat. & Man. Processes, 3(4) (1988) 551.
20. B. Edenhofer, Metall. Mat. Technol., 8 (1976) 421.
21. N.Y. Pehlinvanturk and O.T. Inal, in Ion Nitriding, Proceedings of an Int. Conf. on Ion-Nit., 15-17, Sept. 1986, Cleveland, Ohio; T. Spalvins, Ed., p. 179.
22. K. Ozbaysal and O.T. Inal, in Ion Nitriding, Proc. of an Int. Conf. on Ion-Nit., 15-17, Sept. 1986, Clev., Ohio; T. Spalvins, Ed., p. 97
23. K. Ozbaysal, O.T. Inal and A.D. Romig, Jr., Mat. Sci. & Eng., 78 (1986) 179.

24. Metals Handbook, Vol. 4, Heat Treating, 9th Ed., American Society of Metals, Metals Park, Ohio, 1981, 191.

25. K. Ozbaysal and O.T. Inal, J. Mat. Sci., 21 (1986) 4318.

26. J.W. Christain, Phase Transformation in Metals and Alloys, Pergamon, Oxford, U.K., 1975, 411.

27. P. Grieveson and E.T. Turkdogan, TMS-AIME, 230 (1964) 1604.

28. E.S. Metin and O.T. Inal, in Ion Nitriding, Proc. of an Int. Conf. on Ion-Nit., 15-17, Sept. 1986, Clev., Ohio, T. Spalvins, ed. p. 61.

29. N.R. McDonald and G.R. Wallwork, Oxid. Met., 2(3)(1970) 263.

30. S.H. Avners, Int. to Physical Met., McGraw Hill, N.Y., 1974, 321.

PLASMA NITRIDING: AN ANALYSIS
OF PHYSICO-CHEMICAL MECHANISMS
AT THE PLASMA/SOLID INTERFACE

J. L. Marchand
Sollac-Ledepp
11, Avenue des Tilleuls
BP 11 - 57191 Florange, France

D. Ablitzer, H. Michel, M. Gantois
Laboratoire de Genie Metallurgique
Ecole des Mines-Parc de Saurupt
54042 Nancy Cedex, France

ABSTRACT

In order to know the passage in solution of nitrogen in an alpha iron matrix during plasma nitriding treatment, a model has been proposed which is based on a mass balance of the dissolved nitrogen at the plasma/solid interface. This model takes into account the reactions of vibrational nitrogen $N_2(X,v)$ and atomic nitrogen at the surface, as well as sputtering of this surface by heavy energetic particles. The originality of this approach lies in the possibility of determining the evolution with time of the surface nitrogen concentration which governs diffusion to the heart of the treated material. With knowledge of the evolutionary law of this surface concentration, it was possible to determine nitrogen concentration profile evolutions in time, as well as the increase rate of the iron nitride layers formed at the surface.
Mass variations of the treated material can be foreseen by taking into consideration these physico-chemical mechanisms at the surface (reaction and sputtering).

IN ORDER TO BE ABLE to determine the nitrogen diffusion profile in an alpha iron matrix during nitriding, as well as the possibility of formation of iron nitride layers, it is essential to know the solid surface conditions.
In terms of balance, this translates by the need to know the limit conditions to solve the second FICK equation. If we consider that thermodynamic balance is very quickly reached at the surface, then it is simple, for a pure metal such as iron, to solve the FICK equation with a constant surface concentration.
In his kinetic study of iron carburizing, COUSINOU **(1)** proposed a more realistic representation by introducing the idea of resistance to carbon transfer towards the matrix. Flux of the solid surface carbon is then proportional to the difference between

the concentration at the surface and the concentration which means a thermodynamic balance with the gaseous phase.
RATAJSKI et al. **(2)** has developed a similar model for iron gaseous phase nitriding by introducing an experimentally determined transfer constant.
As done by these two authors for conventional treatments, we have tried to establish a temporal law of surface nitrogen concentration in the case of iron nitriding with molecular nitrogen plasma. With knowledge of this law, it has been possible to determine the evolution of nitrogen concentration profiles in time, as well as the increase rate of surface-formed iron nitride layers, and at the same time to establish a treated material mass variation model.

I-REPRESENTATION OF PLASMA/SOLID INTERFACE PHENOMENA

I.1-Dissolved nitrogen mass balance near solid surface - The three following phenomena must be taken into account in order to express the balance :

a) The atomic nitrogen diffusion from the surface towards the sample center which simply results from the chemical potential gradient and which is independent of the treatment process.

b) The gas-solid reactions which cause the nitrogen to pass from gaseous phase to solution state in the matrix. The plasma species giving rise to these reactions are, as has been seen elsewhere **(3)**, **(4)**, $N_2(X,v)$ species[+] and atomic nitrogen. It must be noted at this point that it is in fact in its atomic form that the nitrogen passes into solution. Consequently, if the $N_2(X,v)$ molecules are reactive, they must be dissociated in approaching, or once on, the surface to be able to react ((PETITJEAN **(5)**).

[+] The $N_2(X,v)$ species are nitrogen molecules at the fundamental electronic state (X) whose vibrational energy is quantified (v).

c) The sputtering of the surface bombarded by fast neutral particles and ions which are accelerated in the dark space surrounding the samples placed at cathodic potential.

The diagram of **Figure 1** summarizes all of the processes which cause surface concentration evolution. The case where the sample is pure iron and the gas is pure nitrogen will be treated.
The variable m^* is introduced and is defined by the following :
$$m^* = m_N / m_{Fe}$$
   - In the case where the matrix is alpha iron, m* is the relation of the dissolved nitrogen mass in the matrix to the mass of iron in the matrix, for an element of volume.

   - For the γ' and ε nitrides, m_N is the nitrogen mass difference in relation to the lower limit of their existance range at the considered temperature, and m_{Fe} the iron mass, for an element of volume.

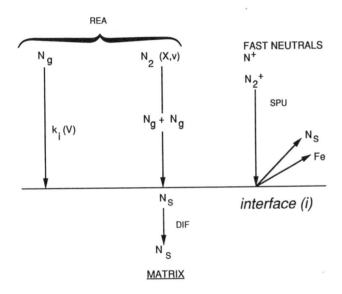

PLASMA

MATRIX

Fig. 1 : Phenomena influencing atomic nitrogen contrentration near the surface of the matrix (i) (i=1 : Fe(α), i=2 : nitride γ', i=3 : nitride ε)

   The sputtering term can easily be defined from this variable.
The surface nitrogen mass balance can be qualitatively written as follows :
$$\frac{dm^*_{s,i}}{dt} = - DIF + REA - SPU \qquad [1]$$
where DIF represents the nitrogen consumption by diffusion towards the heart of the solid, REA represents the nitrogen supplied by reaction of the gas reactive species with the surface, and SPU represents the disappearance of the dissolved nitrogen in the matrix by sputtering of the sample.
Signs s and i indicate that the phenomena take place at the surface of a matrix i.

I.2- Evaluation of the balance terms -
To obtain the surface concentration of dissolved nitrogen, the terms DIF, REA and SPU must now be expressed to integrate relation [1].

I.2.1- DIF term - By making the approximation that the nitrogen diffusion coefficient D_i in matrix i is constant, the second FICK law gives the diffusion term, as a function of the mass percentage, as follows :
$$DIF \approx -D_i \quad (\partial^2 \omega/\partial x^2)_{x=0}$$
ω is the dissolved nitrogen mass percentage at abscisse x, m* and ω are linked together by the simple relation explained in paragraph I.4.

I.2.2- REA term - Taking the hypothesis according to which the reactive species are the atomic nitrogen, and the vibrational nitrogen, the existence of a minimum vibration level above which nitriding is possible must be considered : On one hand, we know that fundamental molecular nitrogen (v=o) is not reactive, on the other hand, optimal nitriding can be expected with atomic nitrogen (v=46). A minimum level of nitriding vibration, v_{min}, thus exists which is dependent on operating parameters and which has been determined in reference (3). Passage of dissolved nitrogen in the matrix can be represented using a kinetic coefficient of chemical reaction. This coefficient will depend on the condition of the molecule (level of vibration v), and on the nature of the solid surface (sign i). The reactions will be taken to be of the first order :
$$N_2 (X,v) == 2 \underline{N}_i \qquad [II]$$
$$N_g + N_g == 2 \underline{N}_i \qquad [IIIa]$$
As atomic nitrogen can be considered as the 46th vibration level (3), relation [IIIa] is therefore strictly equivalent to the following
$$N_2(X,v=46) == 2 \underline{N}_i \qquad [IIIb]$$
REA is thus written as :
$$REA = \sum_{v=v_{min}}^{46} k_i(v) [N_2(X,v)] \qquad [IV]$$
where $K_i(v)$ (in $m^3 s^{-1}$ $mole^{-1}$) is the kinetic coefficient of reactions [II] and [III] ; $N_2(X,v)$/in mole/m^3 is the concentration of the vibrational species v in the gaseous phase near the surface of the solid, and $\underline{N}_i$ (mass percentage) is the concentration of nitrogen dissolved in the matrix i.

I.2.3- SPU term - Surface sputtering, due to bombarding by heavy energetic species (rapid neutral particles and ions), tears off atoms of the matrix as well as atoms of the nitrogen which has already passed into solution.
Considering that, in a first period, the sputtered nitrogen rate is simply going to be proportional to the concentration of nitrogen in solution at the plasma/solid interface, the proportionality coefficient

β_i only depends on the nature of the surface, or :

$$SPU = \beta_i \, m^*{}_{s,i}$$

An expression for β_i is given in appendix.

I.3 - Integration of mass balance

The coefficients $k_i(v)$ being unknown, it is impossible to numerically solve the equation in the manner in which it is posed.
Models of on-surface particle "sticking" exist (LAMPERIERE (6)) which take sputtering into account, and which are based on the hypothesis that term of diffusion of nitrogen towards the solid interior is negligible in the face of the other two terms (REA and SPU). This hypothesis which is accepted here will be verified later in paragraph II.
Thus, if DIF $\langle\!\langle$ REA-SPU, equation [1] can be expressed as follows :

$$\frac{dm^*{}_{s,i}}{dt} = REA-PUL = \sum_{v=v_{min}}^{46} k_i(v) \, [N_2(X,v)] - \beta_i \, m^*{}_{s,i}$$

that is to say : $\dfrac{dm^*{}_{s,i}}{dt} = \alpha_i - \beta_i \, m^*{}_{s,i}$ [V]

in posing : $\alpha_i = \displaystyle\sum_{v=v_{min}}^{46} k_i(v) \, [N_2(X,v)]$

where α_i and β_i (in s^{-1}), once the fixed operating parameters (discharge current and voltage, pressure, etc.) depend only on the nature of the surface. Time zero point (t=0) is chosen either at the beginning of treatment (in this case i=1), or at the beginning of a phase transformation (i=2 and 3). The value of m^* at the plasma/solid interface at time t is given by analytical integration :

$$m^*{}_{i,s}(t) = \frac{\alpha_i}{\beta_i} \, (1 - e^{-\beta_i t}) \qquad [VI]$$

This result is to be compared with that of LAMPERIERE (6) concerning titanium nitride deposit by reactive cathodic sputtering. A material balance at the level of the titanium target surface, permitted his obtaining of an equation which is very similar to that established by us [V]. The variables chosen by LAMPERIERE are different from those that have been used in our work (covering rate replaces mass percentage of nitrogen in solutions, etc.) but representation of phenomena is the same. In the LAMPERIERE model however, a surface covering rate which was a simple function of pressure was considered. In our case, a kinetic coefficient ($k_i(v)$) has been introduced.
Phenomena representation such as we have developed it here, presumes knowledge of reactive species concentrations ($N_2(X,v)$) in gaseous near the surface. COUSINOU (1), for his part, used a surface flux to determine the condition at the limit of the equation of diffusion in the solid. Whatever the chosen approach may be, at the present state of knowledge, it is not possible to avoid an experimental determination of one of these unknowns.

I.4- Nitrogen diffusion in the matrix

An expression [VI] has been obtained which gives the evolution in time of surface concentration of nitrogen in solution. This gives the boundary condition for integration of the second FICK equation. At this point, it is possible to write a differential balance for the nitrogen in solution in the matrix.
The mark zero point (x=0) is chosen at the solid surface. Now seeing that the surface position changes due to sputtering, it is necessary to introduce a convective term into the differential balance which takes plasma/solid interface position change into account. This being, nevertheless, very small, this term has been overlooked. In this way, the second FICK equation is simply written as follows :

$$\partial\omega/\partial t = D_i \, (\partial^2\omega/\partial x^2) \qquad [VII]$$

It continues to be assumed that diffusion coefficient D_i is independent of mass percentage. m^* and ω are linked by the following relations :

for i=1 (Fe((α))) $\omega = m^*/(1+m^*)$ [VIII]

for i=2 (γ') $\omega = (m^*+D)/(1+m^*)$

for i=3 (ε) $\omega = (m^*+B)/(1+m^*)$

where D and B are values of m^* corresponding to the lower limits of nitrides γ' and ε at the considered temperature which can be defined using the Fe-N equilibrium diagram.
In the case of iron(α), considering that $m^* \langle\!\langle 1$ (and thus that $\omega \simeq m^*$), an analytical solution to equation [VII] by use of the condition at the limit given by relation [VI] exists. Given the heaviness of the resulting expression (3) It is preferable to dispose of a numerical solution.
Coefficient α_i and β_i are deduced from the work of IBARRA (7) concerning discharge voltage evolution as a function of surface nature evolution. It is in this way that for a gaseous phase made up of pure nitrogen at 330 Pa pressure that it was possible to estimate the values of α_i and β_i :

for i=1 (Fe(α)) $\alpha_i = 5 \ 10^{-6} s^{-1}$

$\beta_i = 5 \ 10^{-3} s^{-1}$

for i=2 (γ') $\alpha_i = 8 \ 10^{-6} s^{-1}$

$\beta_i = 1 \ 10^{-4} s^{-1}$

Once the solubility limit of the nitrogen in the iron (α) is reached, nitride γ' is formed. The γ'/Fe(α) interface position change is governed by nitrogen conservation equation at this interface :

$$D\gamma' \frac{\partial\omega}{\partial x}\gamma' + D_{Fe(\alpha)} \frac{\partial\omega_{Fe(\alpha)}}{\partial x} = (D - E)\partial Y/\partial t \; [IX]$$

$\omega\gamma'$ and ω_{Fe} are the mass percentages of the nitrogen in solution in γ' and in $Fe(\alpha)$,

Y is the $\gamma'/Fe(\alpha)$ interface abscisse,

D is the masse percentage of the nitrogen in the nitride γ' corresponding to the lower limit of its stability range at the considered temperature,

and E is the solubility limit of the nitrogen in the iron (α).

The diffusion coefficient of the nitrogen in the iron (α) has been extracted from reference (8) :

$$D_{Fe}(\alpha) (823K) = 7.5 \ 10^{-12} \ m^2 s^{-1}.$$

The coefficient of diffusion in nitride γ' is that proposed by MARCINIAK (9) :

$$D\gamma'(823K) = 1.4517 \ 10^{-13} \ m^2 s^{-1}.$$

Discretization (implicit diagram) of equations [VII] and [IX] permit obtaining the concentration profile of the nitrogen in solution in the iron(α) and, for γ', the profile of the difference from the lower limit of its stability range, as well as the increase rate of the γ' layer.

Simulation results are given in **Figures 2a, 2b, 2c** and **2d** for a nitriding experiment with pure iron in a pure nitrogen plasma at 820K.

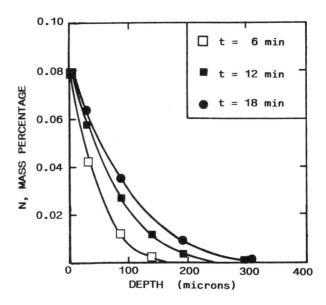

Fig. 2b : Evolution of mass percentage of nitrogen in solution in the iron (α) in time for a treatment at 820K (γ' formed at surface)

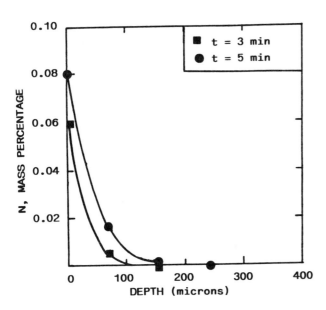

Fig. 2a : Evolution of mass percentage of nitrogen in solution in the iron(α) in time for a treatment at 820K (no nitride formed at the surface)

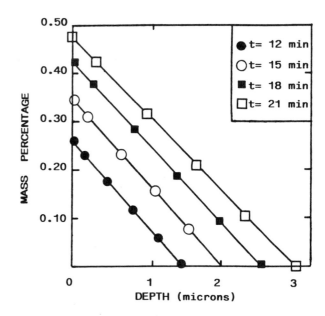

Fig. 2c : Evolution in time of mass percentage in the γ' layer expressed as the difference in relation to lower limit of γ' stability range for a treatment at 820K

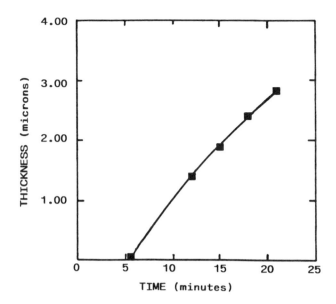

Fig. 2d : γ' layer thickness variation in time. Nitriding at 820K.

Time (minutes)	Nature of the surface	REA α_i (s^{-1})	SPU β_i $m^*_{i,s}(s^{-1})$	DIF (s^{-1})
3	Fe(α)	5×10^{-6}	3×10^{-6}	1.9×10^{-6}
6	Fe(α)	5×10^{-6}	4×10^{-6}	1.3×10^{-6}
12	γ'	8×10^{-6}	2.6×10^{-7}	4×10^{-6}
18	γ'	8×10^{-6}	4.2×10^{-6}	3×10^{-6}

Table I : Comparison of the orders of magnitude of the terms of equation [I] (REA, SPU, DIF)

The hypothesis concerning diffusion rate in relation to surface reaction and sputtering rate has not been verified during the very first instants of treatment. The gradient of the chemical potential in the solid is maximal at the beginning of treatment. Moreover, when γ' has been formed at the surface, diffusion towards the solid center can not be neglected. It should be remembered, nevertheless, that the values of α_i and β_i used for these calculations are estimations which must necessarily be made more precise, and that as a function of the finally retained values, the above conclusion may be modified.

II- DISCUSSION

The experimental results of IBARRA (7) have been used to obtain α_i and β_i by imposing, as it were, an adequation between model and experiment. A conclusion on the validity of this representation can not be reached, even though the general appearance of the profiles obtained seem to be satisfactory. The difficulty lies in determination of parameters α_i and β_i which depend on surface nature (i), but also (in the case of α_i) of the density of vibrational species population at surface of the solid, and thus of all operational parameters (discharge current and voltage, pressure, flowrate, etc.).

An expression has been proposed for α_i elsewhere (3), (10), notably to clearly explain the coefficients of reaction $k_i(v)$ wiht the help of thermodynamic considerations. (β_i is explained in appendix).
The term for the diffusion towards the interior of the solid in equation [I] has been neglected. Without this decision, it was impossible to obtain an expression (analytical or numerical) of the variation of mass percentage at the surface.
Using the obtained mass percentage curves, it is possible to calculate the diffusion term of equation [I] (DIF), and to compare it to the REA (= α_i) term and SPU (= β_i $m^*_{s,i}(t)$) term (**Table I**) for the retained values of α_i and β_i.

III- WEIGHT VARIATION

It has been experimentally noted that, under certain operational conditions, the treated parts lose more weight due to sputtering than they gain due to the nitriding. The heavy energetic species sputter the dissolved nitrogen atoms, but also the much heavier matrix atoms.
In the following, a representation of the phenomena will be proposed which permits determination of for conditions which the total sample mass decreases or increases.

III.1-Overall mass balance

III.1.1- Establishment of mass variation law - Let us consider a pure iron sample whose mass varies in time due to reactions [II] and [III], and due to sputtering of its surface. We will consider, under the surface (A), "thickness e_1" of density ρ_{Fe} defined in appendix.
At time t, the total sample mass is equal to the iron mass plus the mass of the nitrogen in solution :

$$m(t) = m_N(t) + m_{Fe}(t) = (1 + m^*(t)_{s,1}) \cdot m_{Fe}$$

At time t + dt, iron and nitrogen have been sputtered, and the nitrogen has passed into solution :

$$m(t+dt) = m(t) + dm(t)$$
$$= m_N(t) + dm_N(t) + m_{Fe}(t) + dm_{Fe}(t)$$

The total nitrogen mass variation is due to the plasma-solid reactions and the sputtering :

$$dm_N(t) = (\rho_{Fe}\, A\, e_1)(\underbrace{\alpha_1}_{\substack{\text{reaction}\\\text{contribution}}} - \underbrace{\beta_1\, m^*(t)_{s,1}}_{\substack{\text{disappearance}\\\text{by sputtering}}})dt$$

The total iron mass variation is due to the sputtering :

$$dm_{Fe}(t) = -(\rho_{Fe}\, A\, e_1)\,\beta_1\, dt$$

The total mass variation is then deduced :

$$dm(t) = (\rho_{Fe}\, A\, e_1)(\alpha_1 - \beta_1.(1+m^*(t)_{s,1}))dt \qquad [X]$$

Whereas $m^*_{s,i}(t)$ is given by relation [VI] :

$$m^*_{s,1}(t) = \alpha_1/\beta_1\, (1 - e^{-\beta_1 t}) \qquad [XI]$$

As long as no phase transformation has modified the nature of the surface, it can be considered that reaction and sputtering phenomena remain unchanged in time. Even if the increase of the concentration of nitrogen in solution slightly modifies the sputtering volume, the contents remain weak enough that thickness e_1 can be considered as constant in time, befor precipitation of nitride γ'. Indeed, e_1 essentially depends on the mean energy of the particles bombarding the surface.

Integration between times $t=0$ and t of equation [X], taking relation [XI] into account, gives the variation of sample weight :

$$\Delta_1 m(t) = m_{Fe}^o(\alpha_1/\beta_1\, (1 - e^{-\beta_1 t}) - \beta_1 t) \qquad [XII]$$

with

$$m_{Fe}^o = \rho_{Fe}\, A\, e_1$$

In the case where nitride γ' has precipitated to the surface, an identical reasoning gives the total sample mass variation, between time t_1 where nitride γ' is formed at the surface, and time t (t_1 becomes the new time zero point) :

$$\Delta_2 m(t) = m_{\gamma'}^o(\alpha_2/\beta_2(1- e^{-\beta_2 t}) -\beta_2(1+M_N/4M_{Fe})t) \qquad [XIII]$$

with

$$m_{\gamma'}^o = \rho_{\gamma'}\, A\, e_2$$

III.1.2- Part mass evolutions - It is interesting to study variation direction of functions $\Delta_i m(t)$:
that is to say

$$\psi_1(t) = \Delta_1 m(t)/m_{Fe}^o = \alpha_1/\beta_1\, (1 - e^{-\beta_1 t}) -\beta_1 t$$

calculating :

$\psi"_1(t)\ (= -\alpha_1\beta_1\, e^{-\beta_1 t})$, we determine the variation direction of $\psi'_1(= \alpha_1 e^{-\beta_1 t} - \beta_1)$.
Knowing the variations of ψ', we determine the sign of ψ' and the variation direction of ψ.

if $\alpha \leq \beta$

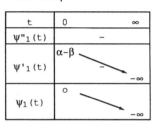

if $\alpha > \beta$

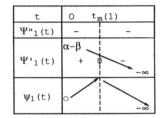

If $\alpha \leq \beta$, that is to say if sputtering is more important than the contribution of nitrogen by reaction, $\psi'_1(t)$ is negative whatever t may may be, and so ψ_1 decreases from 0 to $-\infty$: the sample continually loses weight.
If, on the other hand, $\alpha > \beta$, there is a gain in sample weight, between 0 and $t_m(1) = \ln(\frac{\alpha}{\beta})/\beta$ then its weight decreases.
In the case where nitride γ' is formed, the condition necessary for weight gain (before $t_m(2)$) becomes : $\alpha_2 > \beta_2(1+M_N/4M_{Fe})$.
An interesting case is that for which the nitrogen concentration is maximal at the surface. Diffusion towards the solid center being governed by the concentration gradient (for a given temperature), it is then possible to control the diffusion speed by acting on the temperature (treatment cycles). This maximum nitrogen concentration at the surface is obtained for maximum α_i values and minimum β_i values. We are then in the case of $\alpha_i > \beta_i$.

Expression of β_i (See appendix) shows the direct correlation between sputtering rates, current density (J) and the mean energy of the particles sputtering the surface, via thickness e_i. Unfortunately, it is difficult to modify current density and particle energy, since these parameters are fixed by the necessity of maintaining the parts at a constant temperature. Here once again we find the problem of the coupling of the energetic function of the plasma (which is translated by the sputtering) with its function of reactive species creation (11). It is know today, however, that use of a pulsating discharge current permits separation of these two functions.
It appears necessary to modify the coefficients, that is to say to improve the kinetics of reactions [II] and [III] either by increasing the number of reactive species, or by improving the values of the kinetic coefficients $k_i(v)$. To hope to be able to do this, their expressions as a function of the discharge parameters must be know (3), (10).

IV-SPUTTERING MODEL VALIDATION - We have tried to validate the precedent sample mass variation model. To do this, the samples were weighed before and after treatment, for different operational times. Several tests were carried out with different nitrogen-hydrogen mixtures which permitted controlling the nature of the formed layers.

The curves of Figure 3a show evolutions of sample masses in time for two different cases :

- no nitride is formed at the surface (total pressure = 270 Pa, $H_2-2.5\%N_2$). The sample continually loses weight. This is the case where $\alpha_1 < \beta_1$. Even if a continual weight loss is noted, the part is nitrided (diffusion layer) all the same,

- nitrides γ' and ε are successively formed (total pressure = 270 Pa, $100\%N_2$). A continuous weight gain is noted. The nitrides being quickly formed, the weight loss seen previously, when the surface is $Fe(\alpha)$, is not shown owing to the time intervals separating each measurement.

Samples in a $H_2-20\%N_2$ atmosphere (total pressure = 270 Pa), have also been treated to study the case where a nitride γ' layer ($i=2$) is formed on the surface. The curve of Figure 3b indicates that after a mass loss (higher than 1% of total sample mass), due to surface sputtering when no nitride has yet been formed, the sample undergoes a mass gain (higher than the total mass lost since the start) ; and then after a maximum, the total sample mass decreases. It should be noted that it was verified that the nitride γ' has already been formed after an hour of treatment (first point on the curve).

INTERPRETATION

When no nitride grows on the surface, and only diffusion in the $Fe(\alpha)$ takes place, it has been noted that the sample continuously loses weight **(Figure 3a)**.

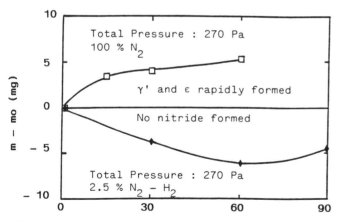

Fig. 3a : Evolution of sample mass for different treatment times at 850 K.

If the plasma nitrogen potential is sufficiently high, the nitride γ' forms after 20 to 30 minutes of treatment. A weight gain is then observed during the time

which corresponds to the nitride growth because sputtering of the surface is less.

When it is experimentally noted that the nitride no longer increases, only diffusion in the $Fe(\alpha)$ seems to evolve. The surface of the nitride γ' layer is still sputtered, but since the layer thickness does not vary, this signifies that the diffusion permits formation of nitride γ' at the $\gamma'/Fe(\alpha)$ interface in a quantity which is equal to that sputtered at the surface.

Lastly, when the γ' layer thickness no longer varies, the sample loses weight due to sputtering of its surface. The only apparent phenomenon is thus the nitrogen diffusion in the $Fe(\alpha)$. Once again it is the situation seen in **Figure 3a**. We have verified that the γ' layer thichness was the same (10 µm) for the last two points of the curve of **Figure 3b**. They do indeed correspond to a mass loss by sputtering.

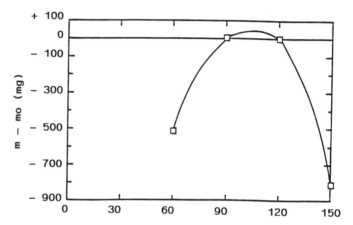

Fig. 3b : Evolution of sample mass when γ' is formed on the surface.

These experimental results and their interpretation confirm model predictions. In order to completely validate this modeling, real values for α_i and β_i must be available.

CONCLUSION

We have determined a law of evolution in time of the nitrogen concentration at the surface of the sample, in the case where the surface is made up of $Fe(\alpha)$, of nitride γ' and of nitride ε. The condition at the limit (surface of the solid) being know, we were able to solve the equations of the diffusion in the matrix, and obtain the nitrogen concentration profiles, as well as the evolution in time of the nitride γ' layer thickness.

The sample mass variation model has foreseen the noted experimental evolutions.

APPENDIX

DETERMINATION OF β_i COEFFICIENTS - In a first period, let us try to make the sputtering coefficient β_1 (case of alpha iron) explicit :

* The number of energetic particles bombarding the surface per unit of time is deduced from the ionic current density J(ions) :

$$J(ion) \; A/q$$

Where A is the sample surface and q is the charge of the ions ($1.6 \; 10^{-19}$ coulombs).
The rapid neutral particles are counted in this expression, since an ion arriving near the surface can either directly reach it, or give rise to a charge transfer generating a rapid neutral. For each ion reaching the cathode, an ion or rapid neutral bombards it (3). These particles arrive on the surface with a mean energy.
The ionic current density is linked to the total current density, J, by the following relations (12) :

$$J(ion) + J(electron) = J$$
$$J(electron) = Y(1) \; J(ion)$$

where $Y(1)$ is the secondary electronic emission coefficient for the iron (i=1). Therefore, we have the following :

$$J(ion) = J / (1 + Y(1))$$

* We can introduce a sputtering rate of the iron by the nitrogen molecules, such as it is classically defined, for an average energy of these particles :

$$\Gamma_i = \frac{\text{number of sputtered iron atoms}}{\text{number of particles bombarding the surface}}$$

The mass of sputtered iron per unit of time then equals :

$$\frac{dm_{Fe}^{SPU}}{dt} = \Gamma_1 \; JA \; M_{Fe}/q\mathcal{N}(1 + Y(1))$$

M_{Fe} being the molar mass of the iron, $\mathcal{N}$ is AVOGADRO's constant.
The sputtered nitrogen mass per unit of time, dm_N^{SPU}/dt, is deduced by considering that the sputtering of a mass m_{Fe} of iron corresponds to the sputtering of a mass $m_{Fe} \; m^*_{s,i}$ of nitrogen (the proportions are maintained) :

$$\frac{dm_N^{SPU}}{dt} = \Gamma_1 \; JA \; M_{Fe} \; m^*_{s,i} \; / \; q\mathcal{N}(1 + Y(1))$$

To obtain an expression of β_i, it is now necessary to ascribe this sputtered nitrogen mass to the matrix iron mass concerned by the sputtering (definition of m*). In this way, the "surface thickness e_1" is introduced. This thickness can be defined as being the relation of the volume of the matrix which is "affected" by the sputtering and the sample surface. It can again be defined by the notion of the material electronic and atomic stopping power (13).

If ρ_{Fe} is noted as the density of iron, the iron mass to be taken into consideration is approximately equal to :

$$\rho_{Fe} \; A \; e_1$$

The variation, due to sputtering of m* per unit of time then equals :

$$\left(\frac{dm^*_{s,i}}{dt}\right)_{SPU} = \Gamma_1 \; J \; M_{Fe} \; m^*_{s,i}/q\mathcal{N}(1+Y(1)) \; \rho_{Fe} \; e_1$$

In this manner, the value of β_1 is obtained :

$$\beta_1 = \Gamma_1 \; J \; M_{Fe}/q\mathcal{N}(1 + Y(1)) \; \rho_{Fe} \; e_1 \qquad [A-I]$$

Identical reasoning for a matrix of γ' or ε leads to the following results :

$$\beta_2 = \Gamma_2 \; J \; M_{\gamma'} \; / \; q\mathcal{N}(1 + Y(2)) \; \rho_{\gamma'} e_2 \qquad [A-II]$$

$$\beta_3 = \Gamma_3 \; J \; M_\varepsilon \; / \; q\mathcal{N}(1 + Y(3)) \; \rho_\varepsilon \; e_3 \qquad [A-III]$$

ACKNOWLEDGE

The authors wish to acknowledge the support of this work by Peugeot S.A..

BIBLIOGRAPHY

(1) **P. COUSINOU**, Thèse de Docteur-Ingénieur de l'I.N.P.L., Nancy, (1977)

(2) **J. RATAJSKI, J. IGNICIUK, J. ZYSK**, 5th Heat Treatment of Materials Busdapest

(3) **J.L. MARCHAND**, Thèse de l'I.N.P.L., Nancy (1988)

(4) **A. RICARD**, Topical Invited Lecture, XVII th I.C.P.I.G., Budapest (1985)
H. MALVOS, A. RICARD, H. MICHEL, M. GANTOIS, 2nd Int. Ion Nitriding/carburizing Conf., 18-20 September 1989, Cincinnati, Ohio (USA)

(5) **L. PETIJEAN**, Thèse de IIIème Cycle, Orsay, (1982)

(6) **LAMPERIERE**, Thèse, Institut de Physique de l'Université de Nantes, (1985)

(7) **C. IBARRA**, Diplôme de thèse de l'INPL, Nancy, (1985)

(8) **J.D. FAST, M.B. VERRIJP**, J. Iron Steel Ind., vol. 176, (1954), p. 24

(9) **A. MARCINIAK**, Surface Engineering, vol. 1, n° 4, (1985), p. 283

(10) **J.L. MARCHAND, H. MICHEL D. ABLITZER, M. GANTOIS, A. RICARD, J. SZEKELY** International Seminar on Plasma Heat Treatment, 21-30 September 1987, Pyc Edition, Paris, p. 85

(11) **M. GANTOIS**, International Seminar on Plasma Heat Treatment, 21-30 September 1987, Pyc Edition, Paris, p. 497

(12) **BROWN** : Basic Data of Plasma Physics (1969)

(13) **N.H. TOLK, J.C. TULLY. W. HEILLAND, C. WHITE** : Inelastic Ion Surface Collisions, Académic Press, Inc., p. 21

ON MICRO-STRUCTURE OF
ION-NITRIDED LAYER

Fengzhao Li
Department of Metallurgical Engineering
and Materials Science
Carnegie-Mellon University
Pittsburgh, Pennsylvania, 15213, USA

Dongsheng Sun, Baorong Zhang
Department of Mechanical Engineering
Shandong Polytechnic University
Jinan, Shandong Province, P.R.C

ON MICRO-STRUCTURE OF
ION-NITRIDED LAYER

Abstract

The micro-structure (fine structure) of ion-nitrided layers in pure iron and structural steels was observed with the thin-film transmission electron microscope. Indexing was made for various diffraction patterns of ion-nitrided compounds. The results indicated that nitrogen atoms were distributed orderly in the nitrides of γ'-Fe_4N and ε-$Fe_{2-3}N$. Micromorphologies of phases and structures of γ', ε and $\varepsilon + \gamma'$ were determined. The authors observed that there were many dislocations, twins and stacking faults in γ' phase. The formation mechanism of crystal defects in γ' phase is also discussed. The relationship between micro-structure and mechanical properties have been studied for different phases and structures.

As a new kind of chemical heat treatment, ion-nitriding has created worldwide interest. Many experts and scholars have carried out research work on the technology and the macro-mechanical properties of ion-nitriding, and have also investigated its mechanism(1). However, ion-nitriding, like any other nitriding method, possesses the characteristics of thin, hard and brittle compound layers, which give rise to problems in further research concerning the fine structure of ion-nitrided layers. As the new technology of ion- nitriding develops, it is very important to investigate the fine structure of the nitrided layer, such as its characteristic, type, morphology together with its distribution of crystal defects and the machanism of accelerating the process of ion-nitriding. Gerardin and others of France(2) using TEM, observed the ion-nitriding layer of pure iron. Their results indicated that there were only twins and stacking faults. Whether or not there are dislocations, their morphology and formation mechanisms need to be observed under TEM.

On the basis of previous work(3), the fine structure and microdefects of ion-nitrided layers in pure iron and structural steels were observed using TEM. The nature and difference of strength-toughness with a single-phase and dual-phase are discussed.

EXPERIMENTAL PROCEDURE

The steels employed in this investigation, are provided in Table I, The compositions of which are in agreement with the respective country standards.

Table I-Analysis of Steels
(% by Weight)

Steel	C	Cr	Mo	Al	S	P
PureIro	0.08					0.004
35CrMo	0.34	0.95	0.32		0.005	0.004
38CrMoA	0.39	1.45	0.32	1.72	0.005	0.004

The three kinds of material were all forged initially and then the pure iron was normalized while the 35CrMo and 38CrMoAl were quenched and high temperature tempered.

The specimens were ion-nitrided with pure ammonia and N_2/H_2 at 1/9 atmosphere, so that an ion-nitrided layer with dual-phases $\varepsilon + \gamma'$, and single phase γ' respectively was obtained. Specimens for TEM were prepared from 0.2-0.3 mm thick slices cut from the ion-nitrided specimens with an electric spark cutting machine, and ground down until the thickness was about 100 μ m. They were thinned chemically, then electropolished with twin-jet polishing unit using 10% perchloric acid and 90% ethanol mixture. The thin foils of the ion-nitrided compound were examined using JEM-800 transmission electron microscope operated at 200 KV.

RESULTS

1. THE METALLOGRAPHY STRUCTURE OF ION-NITRIDED LAYERS

Fig. 1 illustrates the typical metallography structure of the 35CrMo steel and it shows that the nitrided layer consists of a compound layer and a diffusing layer.

2. THE CHARACTERISTIC OF MECHANICAL PROPERTIES OF SINGLE-PHASE (γ') AND DUAL-PHASE ($\varepsilon + \gamma'$)

The former work indicated that the toughness of the single-phase possessed a higher toughness than the dual-phase. The impact strength a_K, bending ,tensile strength and torsional angle of the former were higher than those of the latter. The fatigue limit of the former was obviously superior to that of the later. The nature of all these differences are satisfactorily explained by the fine structure research results in this paper.

3. THE FINE STRUCTURE OF THE SINGLE-PHASE γ'-Fe_4N AND ITS CRYSTAL DEFECTS

3.1 The Fine Structure of Single-phase γ'-Fe_4N

Fig. 2a,b,c show TEM micrograph of single-phase γ'-Fe_4N, its electron diffraction pattern and indexing results.It indicated that besides the diffraction spots of face-centered cubic (FCC) structure (light spots), it may be seen there are also 001, 100 electron diffraction spots, which have resulted from an order distribution on nitrogen in FCC structure (i. e. the diffraction of superstructure). Because there are different scatting coefficients with different atoms in the crystal cell, the order distribution of the atoms will produce lattice translation symmetry, which will result in some new weak superstructure diffraction in this direction. The morphology of the single-phase γ'-Fe_4N (Fig. 2a and 3) shows an equi-axed fine crystal grain.

3.2 The Crystal Defects in γ'-Fe_4N

1) Twin

Using TEM, it was found that there often appeared twin substructure in γ'-Fe_4N. Fig. 4a,b,c are the morphology, electron diffraction pattern of twin and the indexing of twin spots respectively. It may be deduced from the relationship between the spots of the matrix (black spot) and the twin (circle), that the twin plane is (111) and the direction of shear strain is $[11\bar{2}]$ (4).

2) Vacancy loops and dislocations.

Using the TEM, it was found that there were many vacancy loops of varying sizes. Gig. 5 shows the morphology of these kinds of vacancy loops. It can be considered that the vacancies in the ion-nitrided layer are produced by the bombardment of high energy ions. The size of the vacancy loop in Fig,5a is about 1,000 Å, which is equal to 250 atomic spaces.

Because of a continuous ion bombardment, large quantities of point defects appear continuously,and as the groups of point defects become larger, it can collapse into a dislocation loop (5). Fig. 5b & c show the vacancy loops, small dislocation loops, helical dislocations and dislocation networks.

3) The stacking faults in r'-Fe$_4$N.

When the phase r'-Fe$_4$N is formed due to continuous bombardment of high energy ions, the surface of misfit in close packet (111) plane is very easy to produce and thus stacking faults are formed (Fig.5d). In a close packet (111) plane of the FCC structure, a spinel twin will be formed when introducing a stacking fault.

4. THE FINE STRUCTURE AND MORPHOLOGY OF DUAL-PHASE LAYER ($\varepsilon+\gamma'$)

Fig.6 a,b,c are TEM micrographs of dual-phase layer, its electron diffraction pattern and indexing. From Fig. 6a it may be seen that ε phase emerges in the form of parallel strips which are distributed between the equi-axed crystal γ'. The single-phase (r') emerges in the from of equi-axed grain.

DISCUSSION
1. PROCESS OF ACCELERATING ION-NITRIDING BY ION-BOMBARDMENT

In the research work of this paper,it was found that there were high density dislocations in r'-Fe$_4$N, which thus showed that the dislocation mechanism at least plays an important part in the acceleration the nitriding process. Due to Frankel defects produced by high energy ion bombardment,this is an important factor of the accelerating process.

2. THE NATURE OF DIFFERENT PROPERTIES BETWEEN THE SINGLE-PHASE-LAYER (γ') AND DUAL-PHASE LAYER ($\varepsilon+\gamma'$)

The previous work indicated that the toughness and antifatigue characteristic of the nitrided layer of the single-phase was superior to that of the dual-phase(3). First the crystal defects of high density will increase the strength and toughness of the single-phase and will provide greater strength-toughness than the dual-phase. Because the nitrided layer of dual-phase possesses a higher brittleness and ε phase distributes between the equi-axed crystal r', it gives rise to 3-dimensional stresses and facilitates fracture.

CONCLUSIONS

1. Single-phased(γ')has equi-axed grain. In the nitrided layer of dual-phase($\varepsilon + \gamma'$),ε-phase merges in the form of parallel strips which distributes between the equi-axed crystal γ'.

2. Nitrogen atoms in γ'-Fe$_4$N andε-Fe$_{23}$N distribute orderly, i.e. the two phases possess superlattice. There are many dislocations, stacking faults and twin substructures in γ'-Fe$_4$N.

3. The dislocations in the nitrided layer accelerate the nitriding process, at the same time Frankel defects produced by ion bombardment accelerate the speed of nitriding.

4. The abundant substructures in γ'-Fe$_4$N strengthen and toughen the ion-nitrided layer. The irregular distribution of dual-phase in nitrided layer lead to greater brittleness.

REFERENCES

1. O. N. Bytehko, Ю.M. Laghtin, МИTOM, 6, 21-24 (1969) (in russian)
2. D. gerardin, Memoires Scientifiques Revue Metallurgie, 74, 457-467 (1977)

3. F. Z. Li, H. C. Lin and B. R. Zhang, Transactions of Metal Heat Treatment, 2, 48 (1984) (in chinese)
4. K. X. Guo, H. Q. YE and Y. K. WU, "The Applications of Electron Diffraction Patterns in Crystallography", P.260, Science Press, Beijing, China (1983)
5. J. Friedel, "Dislocations", p.92 and P.139 (1964)

Acknowledgment

The authors are indebted to senior Engineers C. Y. Song, X. Y. Li and J. R. Li of the Iron and Steel Research Institute, Beijing, China , for many helpful discussions and to Dr. Bennett (Coventry Polytecnic University, England) for his support and assistance.

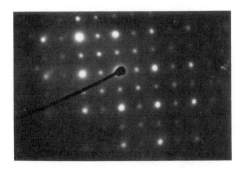

Fig. 2b- γ'-Fe$_4$N diffraction pattern

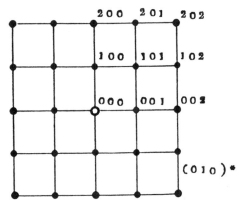

Fig. 2c- γ'-Fe$_4$N Indexing

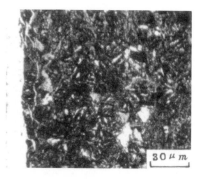

Fig. 1-Ion-nitrided layer of 35CrMo steel

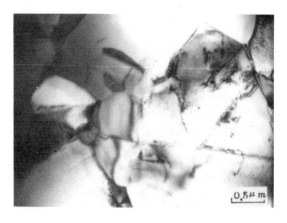

Fig. 2a-TEM microgragh of pure iron γ'-Fe$_4$N

Fig. 3- TEM micrograph of 38CrMoAl steel γ'-Fe$_4$N

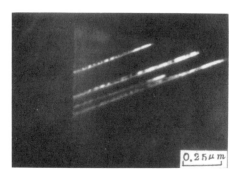

Fig. 4a-The twin in γ'-Fe$_4$N

Fig. 4b-The twin diffraction pattern in γ'Fe$_4$N

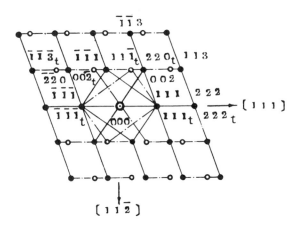

Fig. 4c-Indexing

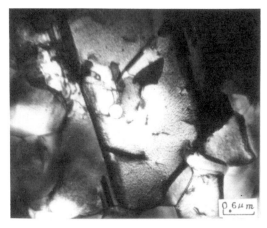

Fig. 5a-Vacancy loops in γ'-Fe$_4$N

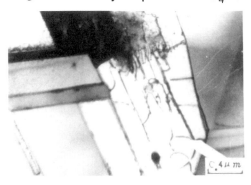

Fig. 5b-Dislocation loops and helical dislocations

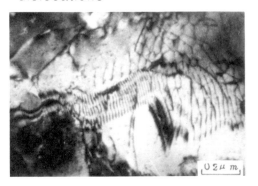

Fig. 5c-Low angle boundaries and dislocation networks

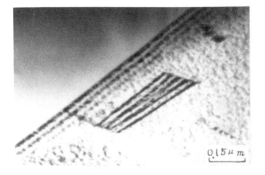

Fig. 5d-Stacking faults

Fig. 6a-TEM micrograph of dual-phase
(ε+ᴦ′) compound layer

Fig. 6b-Electron diffraction pattern

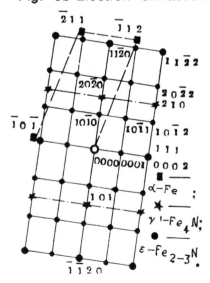

Fig. 6c-Indexing

MATERIALS SELECTION AND PROCESS CONTROL FOR PLASMA DIFFUSION TREATMENT OF PM MATERIALS

Kyong-Tschong Rie, Frank Schnatbaum
Forschungsstelle für plasmatechnologische Produktionsverfahren am
Institut für Schweißtechnik
TU Braunschweig Langer Kamp 8
D-3000 Braunschweig, FRG

Introduction

The increased need for saving raw materials leads to the development of powder metallurgy. Simultaniously the demands for surfaces with higher wear and corrosion resistance increases. At the same time higher fatigue life is required.

PM parts are produced with high precision and with close tolerances. If very close tolerances are demanded the parts have to be machined after sintering. In this case different natures of the surface are found on the machined workpiece and on the as-sintered surface. The nature of the surface influence significantly the layer formation during subsequent diffusion treatment.

To ensure the production of the precision parts by powder metallurgy, processes have to be used which avoid deformation of the parts after diffusion treatment.

Plasma nitriding and plasma nitrocarburizing are most promising surface treatments for improving the wear resistance and the surface hardness of PM structural parts. The great number of independent parameters in plasma diffusion treatment allows close control of the process so that surface layers with defined properties can be produced /1 - 5/.

Experimental

Specimens of three different materials were used for the tests.

Material 1: prealloyed powder
Material 2: powder masteralloy prealloyed with copper, nickel and molybdenium
Material 3: blended elemental technique

The chemical composition of the specimens is shown in table 1, the specimen form and size in fig. 1. As it is shown the specimens were produced from the PM materials with a diameter of 42 mm and a hight of 17 mm. A density between 6.9 and 7.0 g/cm' was obtained. All specimens were sintered in inert gas at 1280 °C, several specimens of the material 1 were sintered in vacuum at 1280 °C. Subsequently all specimens were cut in 4 pieces (fig. 1b). In this way they offer two different surface natures:

1. The surface as-sintered.
2. The surface after machining.
   The machining has led to a surface nature which was almost compact.

The plasma diffusion treatment was performed in two units of laboratory size which work with unpulsed and pulsed direct current. They were discribed in detail by /6, 7/.

The specimens were treated with following parameters:

Temperature: 500 - 600 °C
Gas pressure: 600 Pa
Time: 0,5 - 10 hrs
Gas mixture: see table 2

The specimens were heated by the glow discharge in the treatment gas mixture and cooled slowly in the evacuated furnace after the treatment.

Surface hardness measurements (HV 0,2) and microhardness profiles (HV 0,1) were compiled. The microstructure and the thickness of compound and diffusion layer were examined by standard metallographic techniques, scanning electron microscopy and X-ray diffraction.

Results and Discussion

Compound Layer

The microstructure and the growth rate of the compound layer are influenced by the treatment temperature, the gas mixture and the material composition.

The phase composition of the compound layer was examined by X-ray diffraction. The compound layers of all tested materials were composed of γ'-phase after plasmanitriding. After a treatment time of 10 hrs at 540 °C a small amount of ε-phase arises in the compound layer of material 1. The amount of ε-phase increases by increasing treatment temperature up to 570 °C. After plasma nitrocarburizing with small amounts of carbon in the gas mixture (gas 2) the amount of ε-phase clearly increases while the amount of γ'-phase decreases. After plasma nitrocarburizing 2 hrs at 570 °C with a carbon content of 2 At.-% (gas 4) in the gas mixture, only ε-phase and small amounts of Fe_3C are detected in the compound layer of material 1.

The compound layer of material 3 showed very small amount of ε-phase after 10 hrs plasma nitrocarburizing at 570 °C, 1 At.-% C (gas 3) in the gas mixture. This can be explained by the alloying elements Ni and Cu which retard the diffusion of nitrogen. With increasing amounts of carbon in the gas mixture and with increasing treatment time the formation of ε-phase is favoured.

The thickness of the compound layer of material 3 varies because of the inhomogenous distribution of Ni and Cu at the surface /8/. Ni and Cu hinder the homogenous formation of a compound layer. In material 2 the alloying elements are distributed more homogenously. But the compound layer thickness is still inhomogenous in areas with a high content of Ni and Cu.

The thickness of the compound layer depends on surface structure and material composition. On the machined side of the specimen (material 1) a homogenous compound layer is formed. With increasing carbon content in the gas mixture the thickness of the compound layer increases a little.

In as-sintered specimen thicker compound layer was formed compared to the machined specimen. Additionally to the surface the compound layer is formed on the grain boundaries and on the surface of pores inside the material 1. Little pores in the material near the surface were filled with compound layer (fig. 2). This is due to the higher porosity of the sintered surface. A higher amount of nitrogen penetrates into the material through the pores. The carbon in the gas mixture accelerates the growth rate of the compound layer.

The growth rate of the compound layer of the material 1 is higher than of materials 2 and 3. This is caused by the higher amounts of Ni and Cu in the materials 2 and 3; Ni and Cu retard the diffusion of nitrogen. The growth rate of the compound layer of the materials 2 and 3 is comparable. On the machined specimen the growth rate of the compound layer follows with increasing treatment time a parabolic rate law of $\sqrt{t}$, fig. 3. After a relatively short treatment time the layer thickness on the as-sintered specimen is not exactly definable. For this reason the determination of a rate law was disregarded.

The thickness of the compound layer increases with increasing amount of the ε-phase. This is shown by the comparison of the results obtained by X-ray diffraction measurements and by the layer thickness measurements.

The hardness of the γ'-phase compound layer of the materials 2 and 3 lies between 500 and 600 HK 0,015. With an increasing amount of ε-phase the hardness in the compound layer increases to 720 HK 0,015 and 780 HK 0,015. Higher hardness values up to 880 HK 0,015 were measured in the γ'-ε-mixed phases of the material 1. This can be explained by the higher amount of ε-phase in the compound layer.

The formation of micropores in the compound layer depends on the treatment temperature, the material composition and the surface structure.

In compound layer of the machined spurface of the material 1 less micropores are formed than in the compound layers of the materials 2 and 3.

After a treatment time of 2 hrs no micropores are visible in the compound layers of the as-sintered specimen for all tested materials. With increasing treatment time the porosity of the surface decreases. Small pores on the surface were filled with compound layers and some micropores in the compound layer begin to appear.

Nearly micropore free layers are formed on the machined surface of the material 1 after a treatment in an unpulsed glow discharge without an auxiliary heater at a temperature of 500 °C, fig. 4. The formation of micropores increases with increasing treatment temperature. Fig. 5 shows the compound layer after a treatment at 570 °C. During plasmanitriding in a DC glow discharge without an auxiliary heater a higher treatment temperature is achieved by a higher current density. This causes an increase in the active nitrogen supply at the specimen surface.

According to the theories that the micropore formation is due to a recombination of atomic nitrogen to molecular nitrogen /9/ the amount of micropores might be reduced by lowering the supply of the active nitrogen species. Therefore some experiments were carried out at a temperature of 570 °C with reduced plasma power. The plasma power was reduced by pulsing the plasma with a pulse duration time of 50 μs and a pulse repetition time of 2000 μs. The necessary additional heating was compensated by an auxiliary heater inside the furnace.

The formation of micropores in the compound layer could be reduced markedly by using a pulsed plasma glow discharge. Fig. 6 shows the compound layer on the machined surface of the material 1 after plasma nitrocarburizing at 570 °C in a pulsed glow discharge.

The thickness of the compound layer on the machined specimen was not reduced by using pulsed DC (compare fig. 5 and 6). On the as-sintered specimen a markedly thinner compound layer was formed, fig. 7 (compare fig. 2). This can be explained as follows: To form a compound layer a least amount of compound layer forming species is required which depends on the plasma power and surface nature. If the amount of layer forming species is lowered by reducing the plasma power and at the same time the species can penetrate quickly into the material through pores, the formation of compound layer is reduced. During machining pores on the surface of the specimens disappear largely by deformation. By this means nitrogen could not penetrate into the machined workpiece as quickly as into the as-sintered specimen. The amount of layer forming species was lowered by reducing the plasma power but enough species were present to form a compound layer leading to an equal thickness on the machined surface regardless of difference in power supply. On the as-sintered side the species penetrated quickly through the pores into the material and a thinner compound layer was formed.

Diffusion Layer

Plasmadiffusion treatment leads to a hardness increase in the diffusion layer of the material 1 (fig. 8). This is due to the nitride formation of alloying element chromium /10/. Because of the small amount of the chromium in material 1 chromium nitride was not detected by X-ray diffraction. After a treatment at temperatures above 590 °C a braunite layer is formed under the compound layer.

The hardness in the diffusion layer of the materials 2 and 3 is only slightly higher than that of base materials after plasma nitriding and plasma nitrocarburizing. This is referred to the absence of nitride forming elements such as chromium. While cooling in vacuum iron nitride (Fe_4N) needles are precipitated. After treatment at temperatures above 570 °C small amounts of retained austinite begin to appear.

After treatment above 590 °C a compact layer of retained austenite is formed under the compound layer. While cooling to room temperature the austenite does not transform into ferrite. This is referred to the effect of Ni.

The penetration of nitrogen is influenced by the nature of the surface. The nitriding depth under the as-sintered surface is equal or greater than the nitriding depth unter the machined surface. This is referred to the higher porosity of the as-sintered surface. More nitrogen is able to penetrate through pores and channels into the material. This leads to an increasing nitriding depth.

The nitriding depth increases on the machined workpiece of the material 1 with increasing treatment temperature up to 570 °C. Fig. 9 shows the layer thickness with increasing treatment times following a parabolic rate law of √t.

The nitriding depth depends on the sintering atmosphere likewise. Oxides hinder the diffusion of nitrogen. During vacuum sintering the oxides on the powder particle sufaces are reduced by carbon forming CO and CO_2. Starting from about 1000 ppm oxygen after sintering less then 100 ppm are achieved /11/. By this means the nitrogen penetrates deeper into the material and an increase of the nitriding depth in the machined workpiece of the vacuum sintered specimens was detected in comparison to the specimens sintered under inert gas.

Surface Hardness

Differences in surface hardness of the machined and the as-sintered specimens are not clearly detectable. The scattering band of the measured hardness values is very wide, especially of the as-sintered specimen. This is referred to the higher porosity and the greater roughness of the as-sintered specimen. Therefore, at least 20 hardness measurements were carried out for each side of specimens. Very low or very high hardness values were not considered. Low hardness values are measured if pores beneath the surface break through during hardness measurement. The 20 hardness values have to be in a range of scatterband of ± 50 HV 0,2.

The surface hardness of the material 1 increases significantly (fig. 10) after plasma diffusion treatment. After plasma nitrocarburizing 5 hrs at 570 °C with 1 At.-% C in the gas mixture the surface hardness increases from 150 HV 0,2 to 590 HV 0,2. This can be explained by the increased hardness of the diffusion layer which acts as an effective supporting element for the compound layer.

Materials 2 and 3 show only a slight increase of surface hardness compared to the material 1. After plasma nitrocarburizing 5 hrs at 570 °C with 1 At.-% C in the treatment gas mixture the surface hardness of the material 2 increases from 180 HV 0,2 to 430 HV 0,2, the surface hardness of the material 3 increases from 150 HV 0,2 to 380 HV 0,2. The slight hardness increase is referred to the lower hardness of the diffusion layer (fig. 8).

After plasma nitrocarburizing of the material 1 in a pulsed glow discharge with an auxiliary heater the amount of micropores in the compound layer was reduced. The surface hardness values measured on the machined workpiece after plasma nitrocarburizing in a pulsed and in an unpulsed glow discharge are nearly the same. This was not the case for as-sintered specimen of the material 1.

Plasma nitrocarburizing in pulsed glow discharge increased the surface hardness only up to 540 HV 0,2, while the surface hardness after plasma nitrocarburizing in an unpulsed discharge was 590 HV 0,2. This is referred to a thinner compared layer thickness on the sintered surface after nitriding in pulsed discharge.

Conclusion

This contribution discribes investigations on plasma nitriding and plasma nitrocarburizing of three different sintered steels. The effect of material composition, treatment temperature, treatment time, the nature of the surface and plasma power on the microstructure, hardness and the growth rate of compound and diffusion layer was studied.

Homogenity of the compound layer depends on the nature of the surface and the material composition. Compared to the masteralloyed (material 2) and the mixed materials (material 3) the prealloyed material 1 shows a more homogenous compound layer on the machined side.

The microstructure of the compound layer is influenced by the selection of the treatment parameters. The phase constitution of the compound layer is influenced by the composition of the treatment gas mixture.

Compound layers are nearly micropore free after plasma diffusion treatment at 500 °C. With increasing treatment temperature the amount of micropores in the compound layer increases. The amount of micropores can be reduced by reducing the plasma power.

The compound layer thickness on the machined workpiece is nearly constant for both discharge methods, while the compound layer thickness on the as-sintered specimen decreases by using pulsed discharge. Compound layer hardness up to 880 HK 0.015 was achieved.

After plasma diffusion treatment the nitriding depth of the as-sintered specimen of the material 1 is equal or greater than that of the machined workpiece for both kinds of treatment.

This is referred to the higher porosity of the as-sintered specimen. Through the pores more nitrogen is able to penetrate into the material.

A significant increase in surface hardness and diffusion layer hardness after plasma nitriding and plasma nitrocarburizing can only be achieved by alloying the materials with nitride forming elements. The surface hardness of the chromium alloyed material 1 increases from 150 HV 0,2 to 590 HV 0,2.

Material composition and sintering processes have to be taken into account for the following plasma diffusion treatment.

References

/1/ G. Schlieper, V. Manolache, W. Rembges: Mechanical and Wear Properties of Ionitrided Cr-Mo Sintered Steel; W. A. Kayser, W. J. Huppmann (Edit.): Harizons of Powder Metallurgy, Part I, Proc. of the 1986 Intern. Powder Metallurgy Conference, Düsseldorf 7 - 11 July, 1986, Freiburg: Verlag Schmidt, 1986, pp 491 - 494

/2/ Y. R. Chen: Ion Nitriding of Powder Metal Steels; ASM International Conference of Ion Nitriding, September 15 - 17, Cleveland, USA, 1986, (Abstract)

/3/ M. Ikenaga, K. Akamatsu, K. Kamai, T. Takase: Ion Nitriding of Sintered Steels; Netsu Shari (J. of the Jap. Soc. of Heat Treatment) 20 (1980) April, No. 3, pp 116 - 121

/4/ K.-T. Rie, St. Eisenberg: Surface Treatment of PM Materials by Means of Glow Discharge Plasmas; Intern. Seminar on Plasma Heat Treatment, Senlis, 21 - 23 September, 1987, France, PYC Edition, Paris 1987, pp 241 - 253

/5/ W. Rembges, J. Seyrhammer, R. Klingemann: Möglichkeiten des Einsatzes von plasmanitrocarburierten Sinterbauteilen; Härterei-Techn.-Mitteilungen 43 (1988) No. 6, pp 348 - 353

/6/ T. Lampe: Plasmawärmebehandlung von Eisenwerkstoffen in stickstoff- und kunststoffhaltigen Gasgemischen; VDI-Fortschrittsberichte Nr. 93, Reihe 5 Grund- und Werkstofe, VDI-Verlag, Düsseldorf, 1985

/7/ St. Eisenberg: Plasmadiffusionsbehandlung von Titan und Titanlegierungen; VDI-Fortschrittsberichte Nr. 149, Reihe 5 Grund- und Werkstoffe, VDI-Verlag, Düsseldorf, 1988

/8/ K.-T. Rie, St. Eisenberg: Plasma Nitriding and Plasma Nitrocarburizing of PM Materials; Proc. 5th Int. Conf. Heat Treatment of Materials, Oct. 20 - 24, Budapest, Hungary, 1986, pp 1014 - 1021

/9/ E. J. Mittemeijer, M. van Rooyen, I. Wierszyllowski, H. C. F. Rozendaal, P. F. Colijn: Tempering of Iron-Nitrogen Martensite; Zeitschrift f. Metallkunde, Bd. 74 (1983) No. 7, pp 473 - 483

/10/ V. A. Phillips, A. U. Seyboldt: A Transmission Electron Microscopic Study of some Ion Nitrided Binary Iron Alloys and Steels; Trans. AIME 242 (1968) No. 12, pp 2415 - 2422

/11/ V. Arnhold, R. Wähling: High Performance Components Produced via Vacuum Sintering; Proc. Int. Powder Metallurgy Conf., Orlando, Florida, June 5 - 10, 1988, pp 183 - 195

	C	Cr	Cu	Mn	Mo	Ni	Fe
Material 1	0,3	0,3-0,4	–	<0,2	0,3-0,4	0,3-0,4	rest
Material 2	0,3	–	1,5	–	0,5	4	rest
Material 3	0,3	–	1,5	–	–	5	rest

Tab. 1: Chemical Composition of Materials, wt-%

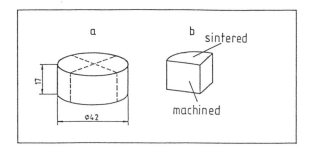

Fig. 1: Shape of the Specimens

Gas mixture 1	80,00 Vol-%	N_2	
	20,00 Vol-%	H_2	
Gas mixture 2	79,80 Vol-%	N_2	
	19,20 Vol-%	H_2	
	1,00 Vol-%	CH_4	$\cong$ 0,5 At-% C
Gas mixture 3	79,94 Vol-%	N_2	
	19,96 Vol-%	H_2	
	2,06 Vol-%	CH_4	$\cong$ 1,0 At-% C
Gas mixture 4	83,10 Vol-%	N_2	
	12,40 Vol-%	H_2	
	4,50 Vol-%	CH_4	$\cong$ 2,0 At-% C

Tab. 2: Compositions of Gas Mixtures

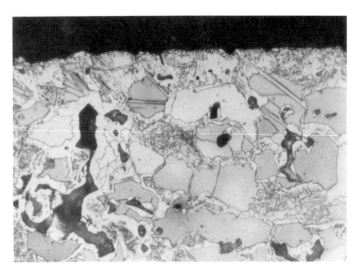

Fig. 2:
Microstructure of Plasma Nitrocarburized
Material 1, 570 °C, 5 h, 500 : 1

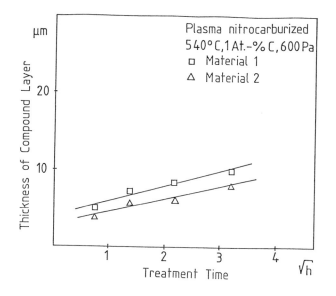

Fig. 3:
Thickness of Compound Layer as Function
of Square Root of Time

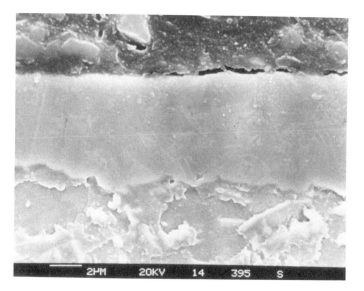

Fig. 4:
Compound Layer of Material 1,
500 °C, 2 h

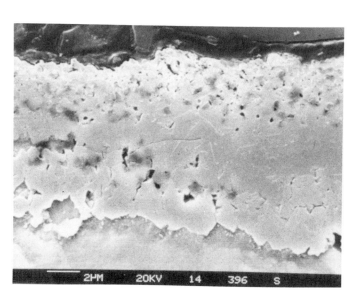

Fig. 5:
Compound Layer of Material 1,
570 °C, 5 h

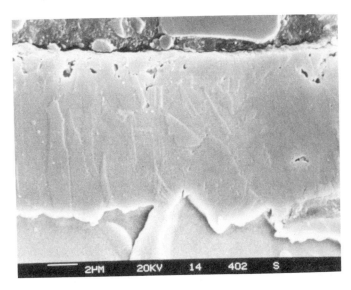

Fig. 6:
Compound Layer of Material 1,
570 °C, 5 h, PD = 50 µs, PR = 2000 µs

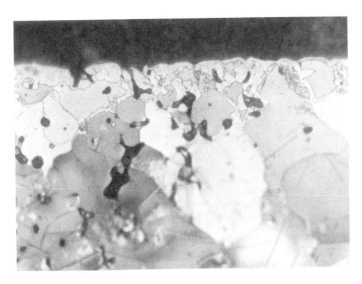

Fig. 7:
Microstructure of Compound Layer
Material 1, Sintered Side,
570 °C, 5 h, PD = 50 µs, PR = 2000 µs

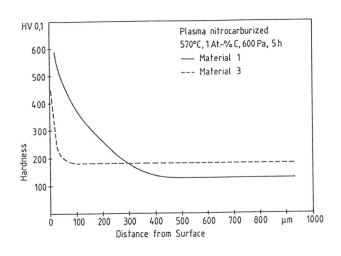

Fig. 8:
Hardness Profiles of Materials 1 and 3

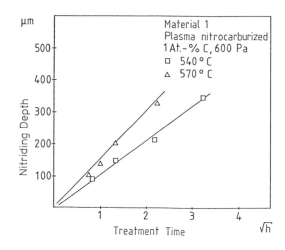

Fig. 9:
Nitriding Depth as Function
of Square Root of Time

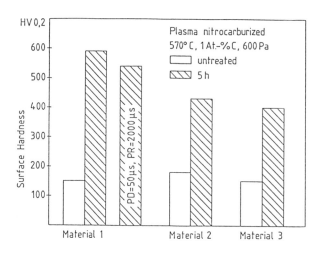

Fig. 10:
Surface Hardness of Plasma
Nitrocarburized Sintered Steel

PRECIPITATION HARDENING OF MARAGE GRADE STEELS DURING ION NITRIDING

Kazim Ozbaysal, Osman T. Inal
Materials & Metallurgical Engineering Department
New Mexico Institute of Mining & Technology
Socorro, New Mexico, 87801, USA

Abstract

Microstructure and properties of several ion nitrided maraging steels (grade 250, 300 and 350) have been studied under varying process conditions, with microhardness-depth correlations, optical microscopy and TEM. The highest nitrided hardness and lowest nitrided case was observed in 350 grade marage steel. In general, the behavior of all three steels in terms of case hardness and case depth were similar. All three steels showed increasing case depths and decreasing surface hardnesses with increasing ion nitriding temperatures and times. An increase in core hardness in all three steels were observed provided that samples were in the quenched condition.

TEM results indicate TiN precipitates in the surface region whereas intermetallic compounds form in the core region. The lath martensitic structure is seen to revert to austenite in the nitrided case.

ALTHOUGH THE ENGINEERING ASPECTS of ion nitriding has been practiced more than two decades, on low and high alloy steels as well as non-ferrous alloys, microstructural analysis of ion nitrided high alloy steels has usually been neglected. This is due to the fact that engineering aspects include mostly the surface hardness and case depth after a reasonable amount of ion nitriding time at a low temperature. Ion nitriding kinetics is usually faster than that of gas nitriding. The reason behind this behavior has recently been delineated by Metin and Inal [1]. Their results on pure iron showed that a metastable nitride, Fe_2N, forms at the surface during ion nitriding. This in turn causes an increase in nitrogen content at the surface. Usually, in gas nitriding, white layer formed consists of Fe_4N, an equilibrium phase, and the concentration at the boundary is fixed by the Fe-N binary phase diagram. With the formation of Fe_2N this nitrogen concentration level is increased at the surface in ion nitriding. The formation of a metastable phase during ion nitriding increases the rate of the process. Since the rate of hardening is faster than that in conventional nitriding , high alloy steels such as stainless steels, tool steels and precipitation hardening steels can be nitrided without affecting the core properties to a large extent. Lower treatment temperatures afforded by ion-nitriding also results in a lower reduction of core hardness. If a precipitation hardening steel is to be nitrided, ion nitriding can be combined with the aging process, provided that the aging conditions are similar to ion nitriding temperature and duration. PH and maraging steels fall into this category. The present study involves characterization of hardening phases during ion nitriding of marage steels, both in the case and in the core regions.

EXPERIMENTAL PROCEDURE

The ion nitriding system used in this study is the same unit used in the previous investigations [2,3]. For all the experiments reported here, the gas mixture was set at 25% volume fraction of nitrogen.

The samples were discs of 2.5 cm diameter and 0.64 cm thick, the surfaces which were ground to 600 grit silicon carbide. The compositions of the steels are given in Table 1. All three grades of marage steels were in the solution treated and quenched condition prior to ion nitriding. The dependence of surface hardness, core hardness and case depth of the three gardes were investigated by ion nitriding between 440 and 520 °C for times 2-10 hrs. Microhardness (VHN) measurements were used to determine the case depth as well as to aid in the characterization of case. For the microhardness-depth correlations, a minimum of three indentations were made at each depth. All microhardness measurements were conducted with a 200 gf load. Samples, which were subjected to microhardness testing, were sectioned, mounted cross sectionally, and polished through 0.06 alumina. Specimens for transmission electron microscopy were prepared by sectioning of the nitrided case with a low speed saw and electropolished in a solution containing 5% perchloric acid and 95% acetic acid. Thin foils of as quenched specimens and specimens from core regions were prepared in a similar way. TEM investigations were carried in a Jeol 100 transmission electron microscope.

Table 1. Composition of the steels investigated; (wt%)

Element	250	300	350
Ni	18.38	18.39	18.04
Co	7.88	9.27	11.82
Ti	0.3	0.59	1.42
Al	0.05	0.12	0.14
C	0.005	0.013	0.014
Fe	Balance	Balance	Balance

RESULTS AND DISCUSSION

After ion nitriding, the steels contained a thin 'compound layer' which was removed prior to hardness testing. In general, 250 and 300 grades received larger case depths than 350 grade. This behavior can be seen in Fig. 1, in the hardness versus depth plot for the samples treated at 480°C for 8 h. The 250 grade showed a greater amount of hardening with low surface hardnesses while 350 grade received the lowest case depth with high surface hardness. The dependence of case depth on time is shown in Fig. 2 for all three grades.

The case depth for all three steels is seen to increase with the square root of time at 480°C. In general, the case depth increases with increasing ion nitriding temperatures for a fixed treatment time due to the increased nitrogen diffusivity. A comparison of the slopes of these curves, shows that nitriding reaction is fastest in 250 and slowest in 350 grade. In general, it was observed that both highest surface hardness and the highest core hardnesses were observed at the lowest ion nitriding temperature, for shorter ion nitriding times. Softening of the matrix was not a problem during the process, since the quenched structure is already soft due to

its low carbon content and hardening in the core region occurs from the precipitation of intermetallic compounds.

Among the three grades, 350 grade received higher surface and core hardnesses than the other two grades. 350 grade contains appreciable amount of titanium, which is a very effective alloying element for hardening by ion nitriding. Titanium nitride is also very stable in steels and does not lose its coherency with matrix even at high temperatures [4]. Therefore, higher titanium content may account for the maintenance of high surface hardness during ion nitriding. This also explains the slower reaction rate observed in 350 grade. Higher titanium content, in turn, leads to higher nitrogen consumption in the case region and decreases the rate of diffusion layer formation.

Fig. 3 shows the 250 grade in the as quenched condition. This grade as well as 300 and 350 grades had a lath martensitic structure prior to ion nitriding. The structure has a high dislocation density prior to ion nitriding and this has an important affect on the strengthening in the core regions during ion nitriding as will be discussed later.

Microstructure of core samples of 250, 300 and 350 grades are shown in Fig. 4. These micrographs are representative of the core regions of the respective samples after ion nitriding at 520°C for 36 hrs. At least four different phases were identified at the core regions. Among these, a bcc phase (Ni_3Ti), and an fcc phase (Ni_3Al). Ni_3Ti and Ni_3Al are the precipitates that causes age hardening in the core in the early stages of ion nitriding. The core regions at this stage (36 hrs of ion nitriding) is relatively soft, due to reversion to austenite and to overaging of intermetallic precipitates.

The case region of 250 grade corresponding to the core region in Fig. 4(a) is shown in Fig. 5. Note that nitrided case is in the overaged condition. Precipitates have grown to a size that make them visible in the bright field and they give ring patterns, as shown in Fig. 5(c). Precipitates were identified as TiN. Dark field image of TiN precipitates in the case region of 250 grade is shown in 5(b). Case regions of 300 and 350 grades are shown in Figs. 6 and 7, respectively. Corresponding core regions were shown in Fig. 4 (b) and (c). Note that, in all cases, the nitride phase is TiN. Case regions from all three grades contained a bcc phase (bcc martensite), an fcc phase (austenite) in addition to TiN.

An examination of the compositions of all three grades shows that titanium content varies from 250 to 350 grade in an increasing order. Considering the amount of titanium in these steels only, 350 grade received higher surface and case hardnesses than the others. The amount of TiN is largest in 350 grade, as shown in Fig. 7 (b). Density of TiN precipitates were less in 250 and 300 grades (i.e, Figs. 5 (b)and 6 (b)). This implies that with an increase in titanium concentration the chemical driving force is increased for TiN precipitation,with an attendant increase in hardness. The hardening phase in the nitrided case of all three steels is TiN.

There is an additional point that can be discussed with respect to nitriding of marage steels. In the core regions, Ni_3Ti precipitated,as well as Ni_3Al. Same samples, in their case region, did not contain any Ni_3Ti but had a dense precipitation of TiN. This means that,when ion nitriding starts, there is a competition between Ni_3Ti and TiN to nucleate. The fact that Ni_3Ti is not observed in the case region implies that activation energy for nucleation of TiN is less and it is favored for nucleation. TiN also consumes most of titanium which would be necessary for the nucleation of Ni_3Ti. Therefore, a firm conclusion regarding to hardening at the nitrided case can be made to suggest that hardening of nitrided region is solely due to TiN. Ni_3Ti or other intermetallics do not contribute to hardening of the case region. In contrast, hardening in core region is due to the precipitation of intermetallics (Fig. 4). It is interesting to note that core region hardens during ion nitriding,provided that the steels are in the as quenched condition.

Formation of austenite was observed in both case and core regions,with x-ray diffraction studies of an earlier investigation [5], and with TEM in this study. Austenite reversion occurs in marage steels after long aging treatment [6,7]. TEM micrographs of both case and core regions are representative of samples after long ion nitriding times (36 hrs). The reason for this long exposure time was to allow precipitate growth so that their identification with electron diffraction would not be complicated. Austenite reversion was observed in the core region of ion nitrided marage steels; basically due to this long treatment time.

There may be additional factors for the formation of austenite in the case region, other than long ion nitriding times. A major reason is the dissolved nitrogen in the case region. The diffusion layer is saturated with nitrogen. Nitrogen, being a very strong austenite stabilizer (about 30 times stronger than nickel [8]), increases the stability range of austenite. Another reason could be the presence of nickel. The fact that no Ni_3Ti was observed in the case region could mean that the nickel that was removed from the martensitic matrix, as nickel-precipitates, is now dissolved in the matrix. Dissolved nickel could also be a factor for

reversion to austenite in the earlier stages of nitriding in the case region.

The last comment to be made with regard to ion nitriding of marage steels is the pre-age condition prior to ion nitriding. In the present experiments, all three grades were in the as quenched condition. This afforded age hardening to occur during nitriding in the core regions of all three grades. If they had been pre-hardened prior to ion nitriding,the core regions would have softened due to over aging and further reversion to austenite would have taken place. The case region on the other hand, would not have hardened fast enough. The reason for this is that now titanium is bound to nickel in Ni_3Ti in the case region and this phase would dissolve to allow TiN to nucleate. This would increase the ion nitriding time and further reversion to austenite. In summary, the as quenched condition is the optimum condition prior to ion nitriding.

CONCLUSIONS

Hardening phase in marage steels during nitriding is TiN. Intermetallic compounds are observed not to be contributing to hardening in the nitrided case. Hardening phases for core regions are Ni_3Ti and Ni_3Al. Softening of the matrix (core region) is due to overaging and reversion to austenite. Reversion to austenite in the case region is due to both long ion nitriding times and dissolved nitrogen.

REFERENCES

1.   E. Metin and O.T. Inal, Material Science, 22, 2783, (1987).

2.   O.T. Inal and C.V. Robino, Thin Solid Films, 95, 195, (1982).

3.   K. Ozbaysal, O.T. Inal, and A.D. Romig, Mat. Sci. and Eng., 78, 179, (1986).

4.   V.A. Philips and A.V. Seybolt, TMS-AIME, 243, 2415, (1968).

5.   K. Ozbaysal and O.T. Inal, On, Ion nitriding - Conference Proceedings, ASM International, Cincinnati, 97, (1986).

6.   D.T. Peters and C.R. Cupp, TMS-AIME, 236, 1420, (1966).

7.   S. Ding, D. Juang and J.W. Morris, Met. Trans., 7A, 637, (1970).

8.   J.C. Lippold, Weld. J., 65, 203, (1986).

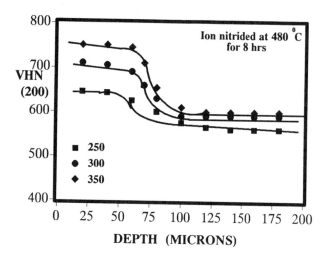

Fig.1 Hardness vs depth characteristics of marage steels ion nitrided at 480 C for 8 hrs

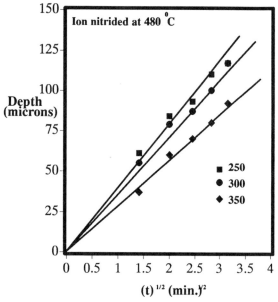

Fig.2 Depth vs square root of time plot for marage steels after ion nitriding at 480 ℃

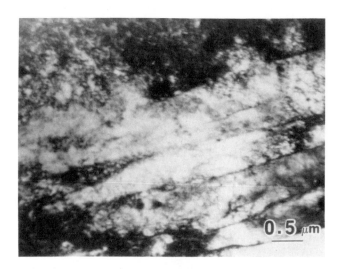

Fig.3 As quenched structure of 250 grade, prior to ion nitriding.

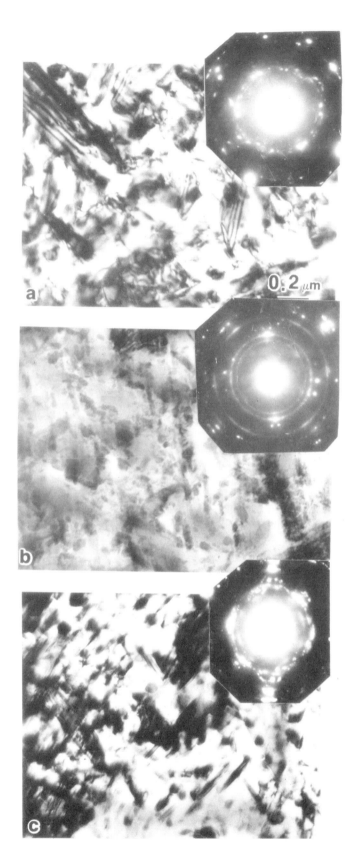

250 Core		
Phase/hkl	d meas.	d th.
fcc/111	2.10	2.07
Ni₃Al/111	2.07	2.07
Ni₃Ti/004	2.07	2.07
bcc/011	2.04	2.03
Ni₃Ti/202	1.95	1.95
Ni₃Al/200	1.88	1.89
fcc/002	1.81	1.79

300 Core		
Phase/hkl	d meas.	d th.
fcc/111	2.10	2.07
Ni₃Al/111	2.07	2.07
Ni₃Ti/004	2.07	2.07
bcc/011	2.04	2.03
Ni₃Ti/202	1.95	1.95
Ni₃Al/200	1.88	1.89
fcc/002	1.81	1.79

350 Core		
Phase/hkl	d meas.	d th.
fcc/111	2.10	2.07
Ni₃Al/111	2.07	2.07
Ni₃Ti/	2.07	2.07
bcc/011	2.04	2.03
Ni₃Ti/	1.95	1.95
Ni₃Al/200	1.88	1.89
fcc/002	1.81	1.79

Fig.4 Core structures of marage steels
after 36 hrs ion nitriding at
520 °C, 250 grade (a),300 grade
(b) and 350 grade (c).

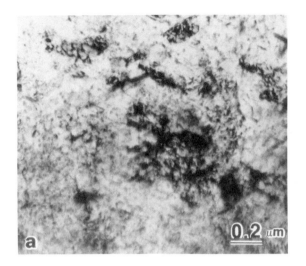

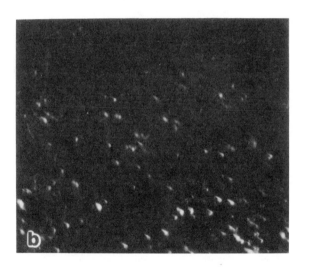

250 Case		
Phase/hkl	$d_{meas.}$	$d_{th.}$
TiN/111	2.45	2.45
TiN/002	2.12	2.12
fcc/111	2.10	2.07
bcc/011	2.04	2.03
fcc/002	1.80	1.79
TiN/022	1.51	1.50
bcc/002	1.45	1.43

Fig. 5 Case structure of 250 grade after ion nitriding at 520 °C for 36 hrs. Bright field (a), dark field from TiN precipitates (b) and diffraction pattern (c)

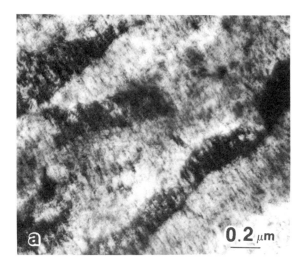

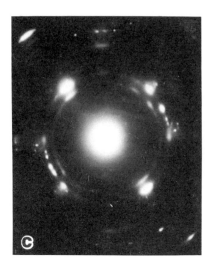

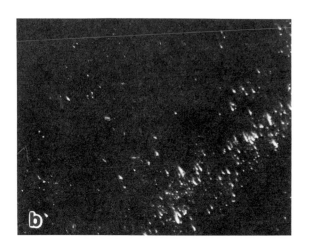

300 Case		
Phase/hkl	d meas.	d th.
TiN/111	2.45	2.45
TiN/002	2.12	2.12
fcc/111	2.10	2.07
bcc/011	2.04	2.03
fcc/002	1.80	1.79
TiN/022	1.51	1.50
bcc/002	1.45	1.43

Fig.6 Case structure of 300 grade after ion nitriding at 520 °C for 36 hrs. Bright field (a), dark field image of TiN precipitates (b) and diffra-pattern (c).

97

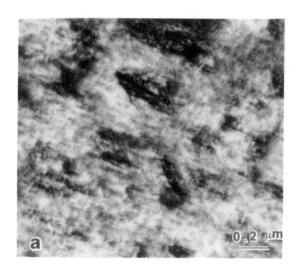

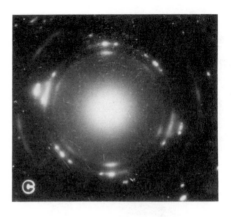

350 Case		
Phase/hkl	d meas.	d th.
TiN/111	2.45	2.45
TiN/002	2.12	2.12
fcc/111	2.10	2.07
bcc/011	2.04	2.03
fcc/002	1.80	1.79
TiN/022	1.51	1.50
bcc/002	1.45	1.43

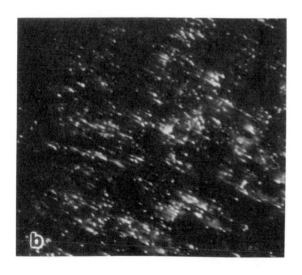

Fig. 7 Case structure of 350 grade after
ion nitriding at 520 °C for 36 hrs.
Bright field (a), dark field of TiN
TiN precipitates (b) and diffraction
pattern (c).

(Ti, Al)N COATING ON PLASMA NITRIDED SURFACES

M. Zlatanović

Faculty of Electrical Engineering
University of Beograd
Beograd, Yugoslavia

Abstract

Two aspects of combined plasma nitriding/coating technology were considered: the deposition of the (Ti,Al)N coating on plasma nitrided samples made of various steel grades and plasma nitriding of partially coated substrates. It has been shown that the excellent adhesion at the coating/nitrided layer interface can be obtained by the careful adjustment of the process parameters. This combined technology can successfully be applied for the surface treatment of cutting and forming tools in order to enhance the tool life and productivity.

PLASMA SURFACE ENGINEERING includes two processes of surface modifications by changing the chemical composition of the surface layer - the diffusion processes and the deposition of coatings. Various mechanical parts made of steel are plasma nitrided or coated in order to get better component performance such as wear and corrosion resistance, lower coefficient of friction, fatigue resistance and resistance to plastic deformation, cavitation erosion and sticking of material, or to serve as a thermal and diffusion barrier, biocompatible and decorative layers etc.

In recent years several authors considered the possibility of successive plasma nitriding and coating treatment in separate or in the same reactor chamber, or to applied plasma nitriding on a partially coated surfaces [1,2,3,4]. In Reference [4] the first industrial application of combined technology has been reported.

Two different zones are formed during nitrogen diffusion in ferrous materials - the compound and the diffusion zone. The compound zone is "responsible" for abrasive, chemical, diffusion and other types of wear, for tribological properties like coefficient of friction, for corrosion and cavitation resistance, sticking of material and other properties related to the contact surface. The fatigue resistance and resistance to plastic deformation of surface layer are related to the diffusion zone properties. In the absence of compound zone the diffusion zone may have also extraordinary tribological properties.

Regarding wear and corrosion resistance and some other properties important for the interaction - contact surface/environment, hard coatings are superior as compared to the compound zone of plasma nitrided layer, but cannot enhance the fatigue resistance and resistance to the plastic deformation of surface layer.

In principle, the extraordinary properties of plasma nitrided layer may be obtained if the compound zone is replaced by a hard coating.

Some properties of combined nitrided/coated layers are presented and plasma nitriding of partially coated surfaces as a possible plasma surface treatment discussed.

PLASMA NITRIDED LAYER AND COATING PROPERTIES

The role of plasma in a various thermochemical and coating surface treatment technologies is not completely understood, but the basic advantage of plasma as an active medium is a very high chemical activity of its components - neutrals and charged particles in different excited states. The plasma/solid surface interaction is of special importance for the surface treatment of various materials. The flux and energy of active species produced in a plasma volume close to the surface to be treated can be controlled by controlling chemical processes in plasma and the electrical potential of the surface relative to plasma potential.

When nitriding iron and various steel grades the reactive diffusion takes place due to a limitted solubility of nitrogen in different phases and a new phase appears directly during the diffusion process and the subsequent slow cooling of the sample. The variety of phases which can be formed due to diffusion of nitrogen or nitrogen and carbon into steel matrix. The crystal structure and lattice parameters of various phases formed onto nitrided surface as a substrate influence the growth and the structure of the deposited hard coatings. Hard coatings like TiN and (Ti,Al)N have a simple B1 NaCl structure, but no iron nitrides or carbonitrides of a simple cubic structure with the lattice parameter close to that of TiN and (Ti,Al)N were found.

A number of hard coatings is available which, in principle, can replace the compound zone of plasma nitrided steel in order to get better wear and corrosion resistance of plasma nitrided steel substrate. Since in some applications plasma nitriding is to be applied after hard coating deposition and subsequent resharpening of selected surfaces, the resistance of hard coating to plasma nitriding may be of importance. This means that a coating has to act as a diffusion barrier against plasma nitriding with high resistance to surface modification in ion nitriding medium.

Since combined plasma nitriding and hard coating deposition techniques are intended to extend the possible applications of coating technology, some additional properties of coatings like performance at elevated temperature may be important in selecting the coating suitable for combined technology.

Metastable (Ti,Al)N polycrystaline films produced by low temperature coating technique are found to be substitutional solid solution of interstitial nitride phase of B1 NaCl crystal structure. Titanium atoms of the TiN lattice are replaced by aluminum atoms and the TiN cell is more and more reduced by increasing the aluminum content, due to smaller size of aluminum atoms compared to titanium. The change of lattice parameter with the relative Al/Ti at.% concentration as given in Ref. [5] can be seen on Fig. 1.

Since the lattice parameter of (Ti,Al)N is closed to that of TiN, it is to be expected that the adhesion of both coatings on the same substrate will be also similar.

Figure 2. illustrates the flank wear of uncoated, TiN and $(Ti_{0.5}Al_{0.5})N$ coated hobs in gear cutting operation. Before the first resharpening of coated tools flank wear was predominant type of wear. After the first resharpening there was no flank wear, but built up edge was the limiting factor of the tool life. Although the tool life was practically the same for both TiN and $(Ti_{0.5}Al_{0.5})N$ coated hobs, the flank wear of $(Ti_{0.5}Al_{0.5})N$ coated hob was always somewhat lower.

From the above mentioned reasons, TiN and (Ti,Al)N coatings were selected to be considered as a possible candidate for combined plasma nitriding and PVD coating technology, and some results concerning the application of (Ti,Al)N coating in this field are presented.

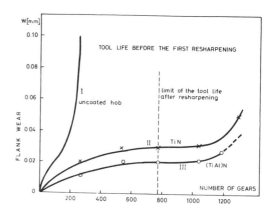

Fig 2. Wear of uncoated (I), TiN coated (II) and $(Ti_{0.5}Al_{0.5})N$ coated (III) hobs; Ref [4]

COATING OF PLASMA NITRIDED SURFACES

New materials suitable for deposition of selected hard coatings can be developed ,or special type of surface treatment applied prior to deposition of coating. Plasma nitriding can improve the fatigue resistance and resistance to plastic deformation of different industrial parts made of various steel grades. Since both properties are related to the structure and composition of the diffusion zone of plasma nitrided layer, the compound zone may be removed and replaced by suitable hard coating of superior surface properties.

The double cathode system described in Ref. [6] designed by Leybold AG was used (Fig.3a.), as well as the additional anode configuration (Fig.3b).

The samples 20 to 25 mm in diameter and 30 mm long were made of different steel grades. Laboratory equipment for plasma nitriding was used in which all the process parameters are

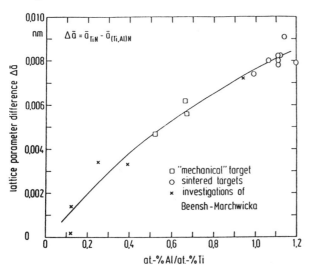

Fig. 1. Change of lattice parameter of the (Ti,Al)N coating

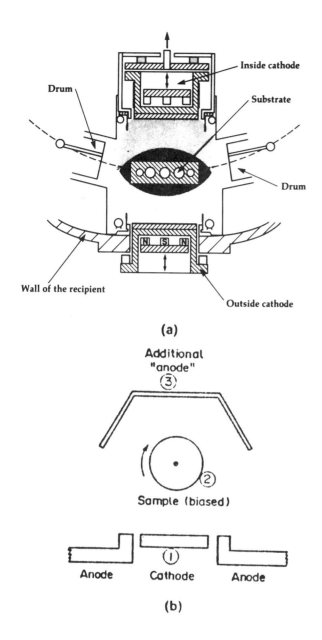

(a)

(b)

Fig. 3.a) Double cathode system b) Additional anode system

Sample No		$L_c |N|$	Plasma nitriding parameters		
			$T | ^oC|$	$N_2 at\%$	$|h|$
4751/	1 2	85 135	480	10	4
B /	1 2	69 71	480	10	6
C /	1 2 3	16 24 15	500	40	1

Table I Conditions of plasma nitriding- steel DIN X38CrMoV51 (AISI H11)

carefully controlled. The working fluid composition was measured and controlled by the thermal mass flowmeters. In all experiments the composition of nitrogen-hydrogen mixture was 10 at.% N_2 and 90 at.% H_2. Relatively low temperature of nitriding and short treatment time were selected in order not to get thick plasma nitrided layer. Before plasma nitriding the samples were degreased. The sample identification number and parameters of plasma nitriding process used in the experiments with the samples made of steel grade DIN X38CrMoV51 (AISI H11) are given in Table I.

Plasma nitrided samples were coated by (Ti,Al)N in a static mode (without the rotation).

The deposition parameters were the following: Deposition rate around 14 μm/h,

pressure 1 Pa, cathode power 6.3 KW, substrate to cathode distance around 60 mm and nitrogen flow 56 sccm. The bias voltage was slightly changed during deposition but in the most part of the process it was -110 V.

The surface microhardness of coating was measured on all samples. The loads used for Vickers measurements were 0.1 and 0.25 N. The surface roughness of coated samples was R_a=0.5-1.2 μm which was nearly the same as that measured on uncoated and plasma nitrided surfaces before deposition of hard coatings.

The thickness of the coatings was measured by a ball grinding method called "calotest" and the critical load was determined by "REVETEST", LSHR produced scratch testing equipment with linearly increasing load acting on the diamond stylus.

Metallographic cross section of sample 4751/2 is given in Figure 4 a). No white layer is visible and the diffusion zone is revealed after etching with 3% nital solution. The stress distribution inside diffusion layer is visible in the form of two sublayers; one of an almost homogeneously stress distribution beneath the compound zone and the other one with continuously decreasing stress concentration towards the base material . This is also visible on the diagram giving the microhardness profile on the sample cross section -Fig.4 b).

Microhardness distribution is excellent concerning plasma nitriding of hot forming tools. As an example of (Ti,Al)N coating deposited on a "white layer" the SEM micrograph of the cross section of samples made of steel grades AISI C 1043 and AISI 4140 are given in Fig. 5. Murakami etching was applied. It can be seen that a compound zone was formed on both samples. A part of (Ti$_{0.5}$Al$_{0.5}$)N coating close to the interface coating/compound zone has a different stress distribution than the upper part of the coating. Some cracks may be seen on the coating interface of the sample AISI C 1043.

A very good adhesion of the (Ti$_{0.5}$Al$_{0.5}$)N coating deposited on plasma nitrided hot working steel DIN X38CrMoV51 together with high oxidation resistance of (Ti$_{0.5}$Al$_{0.5}$)N coating offers a possibility for a combined plasma surface treatment of hot working tools. Some

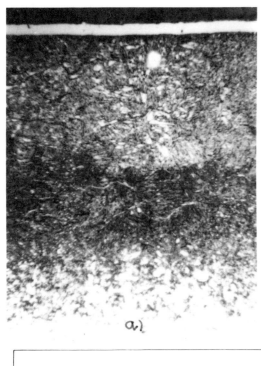

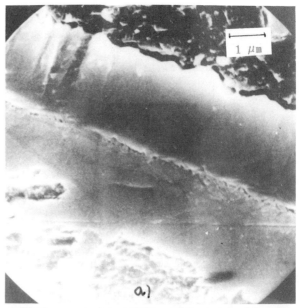

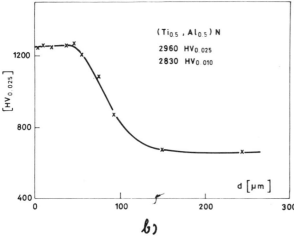

Fig 4. Sample 4751/2 a) Metallographic cross section of ;4b) Microhardness distribution

additional information on the resistance of the $(Ti_{0.5}Al_{0.5})N$ coating to thermal cracks and its thermal expansion coefficient as compared to that of hot working steel substrate are needed.

The structure of plasma nitrided/coated layers is visible from SEM micrograph of broken part of various samples. All coatings have columnar but dance structure.

The influence of substrate surface properties on the $(Ti_{0.5}Al_{0.5})N$ coating adhesion (fig.6.) is clearly visible. In the case of sample 4751/2 (DIN X38CrMoV51) the coating structure is dance and columnar and very good adhesion is expected since almost smooth interface was formed (Fig.7).

By EDX analysis the ratio of Al/Ti in the coating was found to be 50/50 at.%, while Auger

Fig.5. SEM micrograph of cross section of steel sample a) AISI C 1043 b) AISI 4140

spectra have shown the composition $(Ti_{0.5}, Al_{0.5})N$.

PLASMA NITRIDING OF PARTIALLY COATED SURFACES

In practice, some surface areas of industrial parts are to be plasma nitrided and not the others . This can be true also when the deposition technique is used for tool surface treatment. Cutting tools with hard coating have to be resharpened after a certain amount of wear is reached. This resharpening procedure eliminates the surface layer obtained by previous surface treatment completely or

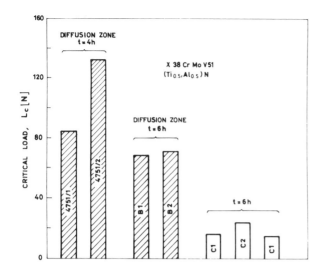

Fig. 6. Critical load of $(Ti_{0.5},Al_{0.5})N$ of steel DIN X38CrMoV51 plasma nitrided at different conditions

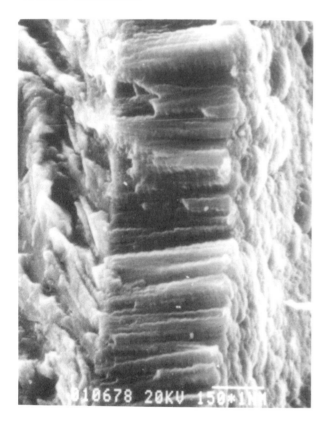

Fig.7. SEM micrograph of fractured cross section of (Ti,Al)N coated substrates; sample 4751/1

partially and to attain again the same tool quality a surface treatment has to be repeated. After a partial removal of a surface layer the recoating can be applied. In some applications it would be more convenient to apply plasma nitriding on partially reground surfaces of previously coated tools than to coat them again.

For this type of an additional plasma surface treatment to be used in practice it is necessary not to have the degradation of the coating properties caused by subsequent plasma nitriding. This means that the influence of plasma nitriding on a hard coating surface properties has to be known.

The influence of plasma nitriding process on the sputtering rate of the (Ti,Al)N coating was tested in a separate experiment with (Ti,Al)N coated high speed steel samples held at the stationary temperature of 380° C during plasma nitriding in a working fluid atmosphere of the composition 25 at.% N_2 and 75at.% H_2 for 12 hours. The deposition process parameters were $P_C=6,3kW$, $f_{N2}=62sccm$ and $U_B=-110V$. The microhardness of coating was $2600HV_{0.05}$ This sample was plasma nitrided at the following conditions: p=2,66mbar, U=800V, I=2A.

The measurements of coating thickness after plasma nitriding were made by calotest and by the direct measurement on the cross sections of the samples. Sputtering rate was found to be $0,01 \mu m/h$ or even lower, which was at the limit of the measurement accuracy, so that the conclusion can be drown that in the case of nitrogen active atmosphere used for plasma nitriding the change of coating thickness can be neglected.

From the experiments it may be concluded that the (Ti,Al)N coating acts as a diffusion barrier for plasma nitriding and that subsequent plasma nitriding does not change significantly the properties of the previously deposited (Ti,Al)N coating.

CONCLUSION

Plasma surface treatment of various steel grades which combines plasma nitriding and deposition of hard coatings has been investigated.

In this combined technology plasma nitriding may be applied prior to coating deposition or partially coated substrates may be plasma nitrided.

When plasma nitriding is used as a previous surface treatment, the adhesion of hard coatings deposited on plasma nitrided layers is the principal problem for the technical applications of combined technology.

It has been shown that the excellent adhesion of (Ti,Al)N coatings deposited on plasma nitrided surface can be obtained by the adjustment of plasma nitriding process parameters.

The formation of compound zone of nitride layer reduces the adhesion of deposited coating. If no compound zone exists the combined nitriding/coating technology my be considered as a nitriding process in which a compound zone is replaced by a hard coating. Such combined layer apart from the excellent tribological properties has in addition good fatigue resistance and the resistance to plastic deformation.

REFERENCES

1. A.S. Korhonen et.al. *2nd Int. Conf. Heat Treatment IFHT*, (1982) 333

2. M. Zlatanović and B. Tomčik, *"Heat Treatment '84"* (1984) 34.1

3. H. Michel, M. Gantois and C.H. Luiten, *"Heat Treatment '84"*, (1984), 1.1

4. M. Zlatanović and W.D.Münz (to be published) Reported in *Plasma Surface Heat Treatment '87*, Senlis, Paris (1987) and *PSE* GarmishPartenkirchen (1988)

5. W D Münz, *J.Vac.Sci.Technol.*, A4(6), (1986) 2717

6. W D Münz, D.Hofmann and K.Hartig, *Thin Solid Films,* 96 (1982) 79

TRIBOLOGICAL PROPERTIES OF ION IMPLANTED
MODEL ORTHODONTIC APPLIANCES

Robert P. Kusy, Stephen W. Andrews
University of North Carolina
Chapel Hill, North Carolina, 27599, USA

ABSTRACT

Four 0.018"X 0.025" arch wire alloys [stainless
steel (S.S.), cobalt-chromium, nickel titanium,
and beta-titanium] were tested against simulated
brackets (1/4" diameter X 1/2" high S.S. cylinders). In addition to control samples, the
polished flats of these cylinders were implanted
with N+, N+/Cr+, N+/C+, C+, Ti+/C+, Ti+/N+, and
Ti+. All were implanted at 2 X 10^17 cm^-2
except Ti+ (4 X 10^17 cm^-2) and Cr+ (3 X 10^17
cm^-2). Quality control was insured by specular
reflectometry, microhardness tests, and Auger
spectroscopy. Using an Instron machine, each
arch wire was drawn at 1 cm/min between the
flats of two cylinders at 34 C in saliva. By
varying the normal forces from 0.2 to 1.0kg, the
frictional forces were measured, and the frictional coefficients (u) were obtained. Kinetic
coefficients (u^k) of wires against the control
S.S. samples measured 0.16, 0.14, 0.24, and
0.31, respectively. Results revealed that, with
few exceptions, the S.S. control cylinders
yielded lower u^k values than the implanted cylinders. Reference to the Auger depth profiles
suggested that optimal ion distributions for
wear resistance were too penetrating for frictional reduction when subjected to low stress,
no wear regimes. Mechanical polishing and laser
annealing were explored as possible remedies.

IN ORTHODONTICS, whether aligning irregular
teeth or closing space after a tooth extraction,
a bracket bonded to a tooth must slide along an
arch wire as forces are transmitted to the
tooth. To obtain a net load of desired magnitude, the frictional forces that develop
between the bracket and the wire must be
overcome by using proportionately greater forces
of activation.

A reduction in the coefficients of friction
(u) between brackets and arch wires would
increase the efficiency of wire movement within
the bracket for tooth alignment and improve
sliding mechanics. Past research was directed
towards producing coatings on different wire
products that reduced u. In 1979, Greenberg and
Kusy [1] coated orthodontic arch wires with a
polymer composite and a polytetrafluoroethylene-
based coating. Respective values for u were
0.073 and 0.028 as compared to 0.16 for uncoated
wires. These preliminary results suggested that
a surface coating could result in at least a 15%
increase in force transmission. The low standard deviations further suggested that the force
transmission could be more reproducible. Unfortunately, the surface coatings tended to crack
on bending, to reduce dimensional tolerances in
the bracket slots, and to stain or peel-off.

Ion implantation alters surface properties
of metals, reducing wear and friction between
contacting surfaces without changing tolerances
[2]. Ion implantation of energetic nitrogen
ions into the surface of Ti-6Al-4V alloy used
for hip and knee replacements significantly reduces wear on both the metal and the ultra-high
molecular weight polyethylene bearing surfaces
[3]. This is explained, in part, in terms of a
much lower u of the ion implanted surfaces.

Surface asperities, which contribute to
friction, can be significantly modified, and
lower values in u can result. In 1973, Hartley
et al. [4] published the first friction tests on
implanted surfaces using a simple slow-speed
sliding apparatus. A variety of ions (Sn, In,
Ag, Pb, Mo) were implanted at fluences in excess
of 10^16 cm^-2 and at energies of 120keV. Macro-
scopic changes in u occurred, and the majority
of implanted ions decreased u. In more recent
years, numerous reports have indicated that
steels implanted with Ti [5-8], with Ti + N [6],
with Ti + C [8-10], and with Cr [11] have decreased u. Additionally, ion implantations with
N and C have demonstrated sharp reductions of
friction and wear for Ti alloys [12].

Because a decrease in friction between
orthodontic brackets and arch wires would lead
to a more efficient, reproducible force transfer

during tooth movement, the effects of implanting various ion species into bracket models and of drawing various wire alloys between them were assessed.

MATERIALS AND METHODS

TEST SPECIMENS - Serving as bracket models, the contact flats of right-hand cylinders (d X h = 1/4" X 1/2") of type 304 stainless steel were polished to a 320 grit finish— a roughness that is typical of orthodontic bracket appliances. Additional cylinders were lapped to a 1 micron finish and used for the quantitative tests. All cylinders were prepared using wet silicon carbide papers and standard metallographic specimen techniques (Table 1).

The arch wires represented the four major orthodontic alloy groups: stainless steel, cobalt-chromium, nickel titanium, and beta-titanium (Table 1). All arch wires were 0.018" X 0.025", except for the B-Ti product, which was 0.017"X 0.025". All wires were used in the as-received condition after ultrasonic cleaning.

PRE-IMPLANTATION TESTS - Prior to any testing or surface modification the surface textures of the arch wires and contact flats were measured using a helium-neon laser reflectometer* [13]. The relative surface reflectance (RSR) of all samples (contact flats and arch wires) were

scanned at an incident angle of 82 degrees as detailed elsewhere [14] to verify that they conformed to the nomogram previously derived. Each sample was scanned at three separate locations to determine an average RSR, that is, the quotient of the reflected intensity from the relatively smooth test surface (I_x) and the incident laser beam (I_o).

ION IMPLANTATION - After this preliminary surface characterization seven ion combinations, energies, and fluences were selected for ion implantation into the contact flats of cylinders (Table 2). Eight 320 grit cylinders for friction testing and two, 1 micron cylinders for quantitative testing were implanted for each ion combination. Unimplanted (UI) specimens provided controls and facilitated comparisons to previous studies.

POST-IMPLANTATION TESTS - To ascertain any changes in surface roughness between pre- and post-implantation, implanted cylinders were randomly selected and their RSR's were evaluated using the laser reflectometer. The new values were compared with the unimplanted values using paired t-tests at the 0.05 level of significance.

Because ion implantation oftentimes enhances the surface hardness, the Knoop hardness number (KHN) was measured* on each control and implanted 1 micron cylinder. Each cylinder was evaluated four times at loads of 2, 5, 10, 15, 20,

---

* Model ML-810, Metrologic Instruments, Inc., Bellmawr, NJ.

* Kentron Microhardness Tester, Kent Cliff Labs, Peekskill, NY.

Table 1 - Materials Evaluated

Contact Flats—

Alloy Type	Code	Nominal Finish	Grinding Sequence
Stainless Steel	S.S.	320 Grit	240 and 320 grit carbide (wet)
Stainless Steel	S.S.	1 micron	320, 400, and 600 grit carbide (wet); then 5 and 1 micron alumina lapping

Arch Wires—

Alloy Type	Code	Size	Configuration	Product
Stainless Steel	S.S.	0.018"X 0.025"	straight wire	Unitek Standard a
Cobalt-Chromium	Co-Cr	0.018"X 0.025"	straight wire	Yellow Elgiloy b
Nickel Titanium	NiTi	0.018"X 0.025"	preformed arch	Nitinol a
Beta-Titanium	B-Ti	0.017"X 0.025"	preformed arch	T.M.A. c

---

a Unitek/3M Corporation, Monrovia, CA.
b Rocky Mountain Orthodontics, Denver, CO.
c Ormco Corporation, Glendora, CA.

Table 2 - Ion Implantation Schedule for
AISI Type 304 Stainless Steel Flats

Ion Specie	Energy (keV)	Fluence X 10^{-17} (cm^{-2})	
N+	60	2	a
N+/Cr+	60/120	2/3	b/b
N+/C+	60/60	2/2	a/c
C+	60	2	c
Ti+/C+	170/60	4/2	b/b
Ti+/N+	170/60	4/2	b/b
Ti+	170	4	b

a North Carolina State University, Raleigh, NC.
b Spire Corporation, Bedford, MA.
c Oak Ridge National Laboratory, Oak Ridge, TN.

30, 40, and 50g.

One 1 micron lapped right-hand cylinder of each implanted species was also evaluated by Auger electron spectroscopy (AES)*. This analysis provided an in-depth compositional profile of iron as well as the specific implanted ions aforementioned (Table 2).

FRICTION TESTS - The coefficients of static (u^s) and kinetic (u^k) friction were measured using an Instron Universal Testing Machine**, a friction apparatus [1], and an environmental test chamber. In this friction apparatus (Fig. 1) a spring transmits the normal force via a moveable piston to the flats, which are in contact with the arch wire surfaces. The magnitude of the normal force (N) was measured by means of a calibrated force transducer (T^N). The millivolt output was monitored with a high impedance recorder. The flats were drawn along the gripped arch wire, which was secured to the machine's load cell transducer (T^P), by the screw driven Instron at a rate of 1cm/min. The millivolt signal from the drawing force transducer was digitally stored at ten data points per second.

After testing at 34 C in one of the investigator's saliva (S.W.A.), the digital data was transferred to a microcomputer for analysis. From the drawing force (P) versus distance trace for each combination, the frictional force (f = P/2) was retrieved. One software program evaluated u^s and u^k for all five normal force loads (200, 400, 600, 800, and 1000g) by positioning cursors on the screen and entering the millivolt signals of the maximum static and the normal

forces. Another program determined the values of u and their respective correlation coefficients from the linear regressions of the five normal force and frictional force coordinates. A third software package plotted the linear regressions for each u^s and u^k.

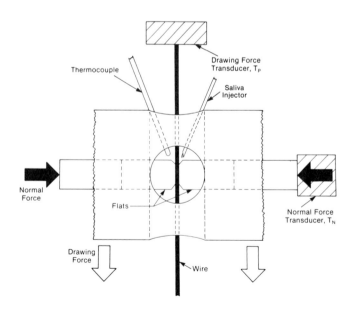

Fig. 1 - Diagram of the frictional testing apparatus in which an arch wire extends vertically between the two loaded flats of the sliding cylinders.

---

*   JAMP-30 Auger Microprobe, JEOL, Tokyo, Japan.
** A modified Model TTCM with 500kg load cell, Instron Corporation, Canton, MA.

After ion implantation, visual inspection of all cylinders did not indicate any obvious surface alterations. Paired t-tests for surface roughness of the implanted samples (Table 3) revealed significant differences ($p < 0.01$) between only the pre- and post-implanted C+ and C+/N+ surfaces.

Microhardness measurements (Fig. 2) increased in magnitude for the N+ and N+/Cr+ implanted cylinders, with the greatest difference seen for loads of 5g or less. The other five ion implanted combinations decreased in hardness at loads less than 5g, with generally no difference at higher loads. These measurements at the lighter loads provided a more realistic picture of surface hardness changes, although even at these loads the indenter readily penetrated the implanted layer.

The AES of the seven ion implantations (Figs. 3 and 4) revealed the presence of each of the implanted ions within the immediate subsurface. In all cases, iron was the primary constituent of the bulk material. In all Ti+ implantations a decrease in the Fe intensity was observed whenever the Ti+ intensity peaked. For all implantations, except carbon, the surface ion concentrations gradually increased to maximum levels at depths ranging from 35nm to 200nm and then tapered off towards zero by a depth of 300nm. For two of the carbon implantations (C+ and N+/C+) the carbon concentration was abnormally high initially and then decreased sharply, taking on a Gaussian distribution pattern similar to the other implantations.

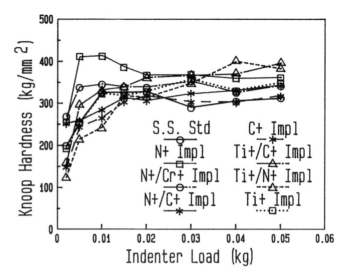

Fig. 2 – Results of the Knoop microhardness test for the seven implantations.

Thirty-two permutations of arch wires and flats were tested, with each yielding f-N plots of the static and kinetic frictional forces (cf Appendix, Figs. i-viii), as obtained from the individual force-distance traces. Without discarding any measurements, the sixty-four values of either u^s or u^k are summarized along with their correlation coefficients in Table 4. The unimplanted results corroborated measurements reported previously [15]. Regardless of the surface, either the S.S. or the Co-Cr arch wires demonstrated the lowest u; NiTi was intermedi-

Table 3 – Relative Surface Reflectance (RSR)

Ion Specie	RSR a Pre-Implantation	RSR a Post-Implantation	p-value b
N+	0.41 +/- 0.02	0.43 +/- 0.04	NS
N+/Cr+	0.39 +/- 0.03	0.40 +/- 0.03	NS
N+/C+	0.42 +/- 0.02	0.43 +/- 0.02	<0.01
C+	0.42 +/- 0.02	0.43 +/- 0.02	<0.01
Ti+/C+	0.41 +/- 0.03	0.42 +/- 0.03	NS
Ti+/N+	0.45 +/- 0.02	0.45 +/- 0.02	NS
Ti+	0.42 +/- 0.02	0.42 +/- 0.03	NS

a The mean RSR +/- one standard deviation of at least eight flats were determined from the relative power (Ix/Io) measurements at an incident angle of 82 degrees.
b Student's t-test of paired data in which NS denotes no significant difference between pre- and post-implantation RSR's.

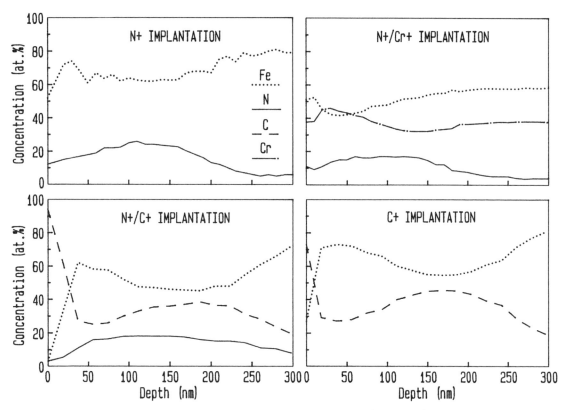

Fig. 3 – Analyses of Auger electron spectroscopy for N+, N+/Cr+, N+/C+, and C+ implantations.

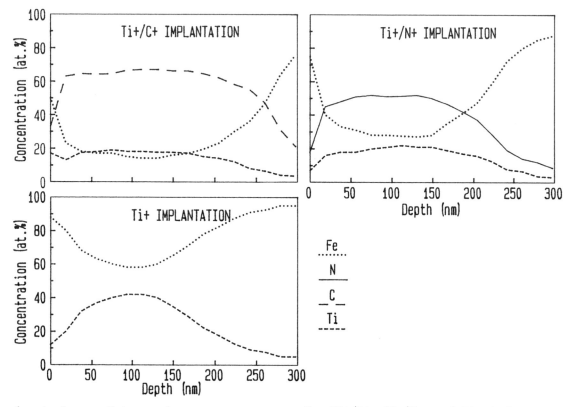

Fig. 4 – Analyses of Auger electron spectroscopy for Ti+/C+, Ti+/N+, and Ti+ implantations.

ate; and B-Ti was the highest. With few exceptions, the S.S. control flats outperformed the implanted surfaces. Generally, any improvements of the implanted cylinders were marginal at best.

The values of u^k for unmodified S.S. measured 0.16, 0.14, 0.24, and 0.31 against S.S., Co-Cr, NiTi, and B-Ti arch wires, respectively. The u^k for the various arch wires against implanted cylinders ranged as follows: for S.S. from 0.12 (N+) to 0.30 (Ti+/N+); for Co-Cr from 0.14 (N+) to 0.27 (Ti+/C+); for NiTi from 0.22 (N+/C+) to 0.39 (Ti+/N+); and for B-Ti from 0.22 (N+/C+) to 0.59 (Ti+/C+).

DISCUSSION

RELATIVE SURFACE ROUGHNESS - Although the RSR measurements were made to insure uniformity among tested specimens, differences between pre- and post-implantation were also ascertained. No apparent change was seen visually after implantations. Finding no significant differences between pre- and post-implantation was expected (Table 3), since no changes to the surface is one advantage of this surface modification. The differences seen with the C+ and N+/C+ were unexpected, since these surfaces are more reflectant, possibly a result of surface contamination with carbon.

KNOOP HARDNESS NUMBERS (KHN) - The indentations made with the Knoop hardness tester penetrated through the implanted layers even at light normal loads, as these layers are of submicron thickness. The depth of penetration of the Knoop tester was calculated from the geometry of the diamond pyramid indenter. Given that the indenter makes an impression whose depth is 1/30th of the long diagonal, the shallowest depth of penetration was calculated for a 2g load on the N+ implanted cylinder. This depth approximately equaled 350nm, a value well beyond that of any implanted layer. Therefore, only an average hardness was measured as a function of normal loads, which with decreasing load converged towards the value of the implanted layer. Consequently in the 15-20g load range, differences between implanted and unimplanted specimens should not be apparent.

As the loads of 5g or less were employed, the N+ and the N+/Cr+ implants increased in hardness in agreement with Iwaki's findings [11]. Using ultra-light loads, Hutchings et al. [16] also found that the relative hardness of N+ implanted type 304 stainless steel increased 1.25 times, a value comparable to our conventional Knoop values. Future hardness testing will employ an ultra-light load indenter to more accurately ascertain the hardness of the implanted layers without penetration.

AUGER ELECTRON SPECTROSCOPY (AES) - As seen in Figs. 3 and 4, all ion concentration versus range distributions were skewed toward the surface. During each measurement, those ions that were implanted into the cylinder and iron were transformed from raw intensity to atomic percent. Each analysis shows that the maximum implantation

concentration is some distance within the surface and suggests that the actual contact during this so-called "light contact stress, no abrasive wear friction testing" may not be against the ion-rich implanted surface at all.

An inverse relationship holds for the presence of iron versus the implanted Ti+ ions. Before implantation the trace of iron should be a constant value. However, in each profile that includes Ti+, the iron value drops-off as the Ti+ concentration increases (Fig. 4). Once the peak concentration of Ti+ is passed, the iron returns to its original value. This finding may be a "matrix effect" in which the Auger electrons are not readily emitted from the iron by the Ti+ during analysis.

The analyses of the C+ and N+/C+ implantations demonstrate additional C+ on the surface as seen by the sharp drop in concentration from the surface to at most 50nm (Fig. 3). This finding suggests carbon contamination resulting from vacuum carburization; if this contamination had occurred during the other implantations, it was not analyzed.

Two additional findings were discovered in reviewing the AES studies. The Ti+/N+ implantation reveals more than double the N+ concentration compared to the other three N+ implants (Figs. 3 and 4). During the Auger analysis, Ti+ produces several peaks one of which is on the same magnitude as the N+ peak. Thus, the measurements listed for N+ in the Ti+/N+ implantation included the values for Ti+ as well as N+. Another anomaly of the Auger analyses was the large concentration of C+ found for T+/C+ (Fig. 4) as compared to C+ or N+/C+ (Fig. 3). Perhaps this difference was attributable to the implantation facility (Table 2).

FRICTION TESTS - The trends encountered among the unimplanted arch wire-contact flat combinations generally followed those that have been previously documented [15,17]: stainless steel provided the lowest u's, followed by cobalt-chromium, nickel titanium, and beta-titanium. Although in these previous studies u's generally exceeded u^k, in the present studies the opposite was observed, regardless of the arch wire alloy. With few exceptions, no improvements were made over the unimplanted controls.

During earlier friction studies, cold welding of the B-Ti arch wire material to the S.S. flats was shown to be one cause of high u's [18]. Although surface modification might prevent such adhesion and stick-slip from occurring, only the N+/C+ implantation on B-Ti improved the u^k from 0.31 to 0.22 (Table 4). No other significant improvement was seen for u^k in this arch wire group. Further examination of Table 4 reveals that any arch wire tested against a Ti+ implantation, whether alone or in combination with another ion, yielded higher u's than the unimplanted couple. In the present study, the presence of Ti+ in the arch wire or the flats increased the coefficients.

In 1983, Pethica et al. [19] observed that

u´s were reduced on the hardened surfaces of implanted materials. More specifically, the reductions in u after Ti+/N+ [6] or Ti+/C+ [8-10] implantations were attributed to the formation of an amorphous layer as well as the formation of nitrides and carbides. These earlier friction reduction results may have been due to carbide contamination on the surface of the implanted materials [20,21]. Studies conducted by Follstaedt and Meyers [22] showed that the amount of residual carbon in the implantation chamber is often sufficient to cause carbon amorphortization in instances where only titanium ions are later accelerated. The Sandia group´s studies suggest that the carbon contamination may have acted as a lubricant providing a film coating.

The values of u´s and u^k in air for unlubricated stainless steel span a wide variation: Bowden and Tabor [23] reported u´s equal to 0.60; the American Institute of Physics Handbook [24] listed u´s and u^k as 0.39 and 0.31, respectively; Nordling and Oosterman [25] quoted the ranges to be 0.15 to 0.30 and 0.15 to 0.20, respective-

ly. These variations may be, in part, a result of the differences in the testing methods. For example, Pope et al. [26] reported the u^k for S.S. in air to improve from 0.50 to 0.30 after implantation with Ti+/C+ versus 0.16 and 0.25 in the present work. Their testing was accomplished using only a 33g normal load for a pin (hemispherical radius of 0.79mm) on a plate. The Hertzian stress at the tip of the pin was tremendous (approximately 3.5 times the yield strength of the unimplanted S.S.), when compared to the stresses that were applied at the arch wire-flat interface. Although the u´s are usually regarded as independent of contact area [27], the concentration of force at a single point apparently promoted higher coefficients— perhaps because of plowing.

In the present friction testing system several factors might have limited the actual contact of the ion-implanted layer with the arch wire surface. Unlike previous ion implantation friction studies, which were conducted under dry conditions, natural saliva bathed the materials

Table 4 - Summary of the Coefficients of Friction for
Ion Implanted Contact Flat/Arch Wire Combinations
(contact velocity = 1 cm/min in human saliva at 34 C)

Ion Specie(s) in Stainless Steel Contact Flats	Arch Wire Alloy							
	S.S.		Co-Cr		NiTi		B-Ti	
Unimplanted (UI)	0.18 a	0.16 b	0.12	0.14	0.34	0.24	0.36	0.31
	0.99 c	0.95 d	0.99	0.99	0.99	0.96	0.98	0.99
N+	0.15	0.12	0.17	0.14	0.21	0.32	0.37	0.31
	0.93	0.92	0.98	0.95	0.98	0.99	0.96	0.97
N+/Cr+	0.25	0.26	0.29	0.23	0.38	0.28	0.32	0.33
	0.94	0.98	0.97	0.96	0.98	0.99	0.90	0.95
N+/C+	0.22	0.24	0.25	0.27	0.32	0.22	0.29	0.22
	0.99	0.98	0.98	0.98	0.92	0.99	0.96	0.93
C+	0.16	0.17	0.25	0.27	0.41	0.31	0.27	0.35
	0.96	0.98	0.98	0.97	0.97	0.95	0.95	0.92
Ti+/C+	0.25	0.25	0.26	0.27	0.38	0.29	0.59	0.59
	0.96	0.98	0.99	0.99	0.95	0.99	0.98	0.96
Ti+/N+	0.26	0.30	0.28	0.26	0.37	0.39	0.60	0.33
	0.95	0.98	0.98	0.98	0.94	0.94	0.98	0.98
Ti+	0.23	0.19	0.16	0.16	0.21	0.24	0.47	0.57
	0.90	0.96	0.88	0.95	0.96	0.97	0.88	0.94

a Coefficients of static friction based on the linear regression of Appendix, Figs. i-viii.
b Coefficients of kinetic friction based on the linear regression of Appendix, Figs. i-viii.
c Correlation coefficients for the values of the coefficients of static friction (p<=0.05).
d Correlation coefficients for the values of the coefficients of kinetic friction (p<=0.05).

to simulate oral conditions. In one study [28], artificial saliva increased the coefficients of friction for S.S., B-Ti, and NiTi compared to dry conditions. Other friction results under wet conditions were less decisive [29,30]. The u^k obtained for the arch wire-contact flat combinations were very comparable to earlier results [15] (0.14, 0.16, 0.33, and 0.35 for S.S., Co-Cr, NiTi, and B-Ti, respectively) for 0.021"X 0.025" arch wires drawn through the 0.022" slot of S.S. brackets in a dry environment. Although the boundary layer that was created by the saliva may have prevented any contact between the arch wires and the ion implanted layers, there is no strong evidence between the dry and wet experiments to support the contention that saliva acts as a lubricant or an adhesive. Consequently, saliva was ruled out as the cause of the present differences between unimplanted and modified appliances.

Another factor that might have influenced the u's between the ion-rich layer of the flats and the arch wire was the relatively low loads applied to the arch wire-contact flat combinations. Among tests that were run by other investigators within our normal load range of 200-1000g, Pope et al. [9] has reported a 50 percent reduction in u^k for loads less than 600g. In comparing the individual u's that were determined for the implanted cylinders at loads of 200 and 400g, the same values were obtained as the overall u^k for the load range of 200-1000g. Thus, the normal loads cannot justify the higher relative values of coefficients that were found for the implanted over the unimplanted couples.

The factor that caused the discrepancy between unimplanted and implanted results was the normal load in combination with the depth of the implantation. The AES traces of Figs. 3 and 4 show that the area of peak implanted ion concentration was some distance subsurface. In the present experimental design, only five passes were made along each implanted cylinder face with no assurance that the wire would ever retrace the same track on a test cylinder. Thus, sufficient surface wear to reach this zone is doubtful. Perhaps if we had used a hardened steel ball slider [5,6], we would have reported improvements in u^k for Ti+ and Ti+/N+ implantations over unimplanted type 304 steel.

In an attempt to draw an arch wire in contact with the maximal implanted levels, the Ti+/C+ implanted cylinders were lapped down 200nm using 1 micron alumina. Prior to lapping, the implanted surface was indented, and the length of the long diagonal recorded. Based on the geometry of the diamond indenter, the length of the long diagonal was reduced about twelve filars after just six seconds of lapping. This surface removal was verified by AES (Fig. 5). When compared with the original post-implantation depth profile (Fig. 4), it demonstrates a shift of the Ti+ and C+ curves towards the surface. The friction testing was repeated only with a stainless steel arch wire, chosen for its consistent performance in previous testing. The S.S. arch

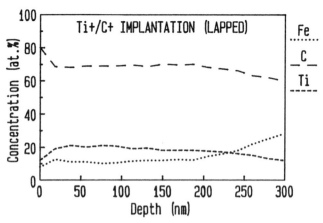

Fig. 5 - Analyses of Auger electron spectroscopy after the flats of Ti+/C+ implanted S.S. cylinders were lapped (cf Fig. 4).

wires were drawn through two different sets of lapped Ti+/C+ implanted cylinders. When the linear regression plots of this data (Fig. 6) were compared to the original Ti+/C+ data (Appendix, Fig. vi), the new u's and u^k equalled 0.21 and 0.19, respectively. These values represented an improvement over the original testing (0.25 and 0.25, respectively), but were still higher than the unimplanted S.S. flats of the control cylinders (0.18 and 0.16, respectively).

In another attempt to obtain the benefits of the implanted layers, the implanted surfaces were annealed with a XeCl laser [31]. After the initial friction tests with the Ti+/N+ implanted contact flats, new S.S. arch wires were drawn between pairs of the laser-annealed surfaces. From the regression plots of the Ti+/N+ implanted (Appendix, Fig. vii) and the laser-annealed experiment (Fig. 7), the values obtained for u's and u^k decreased from 0.26 and 0.30 to 0.21 and 0.21, respectively. Once again, in neither case were the S.S. control values (0.18 and 0.16,

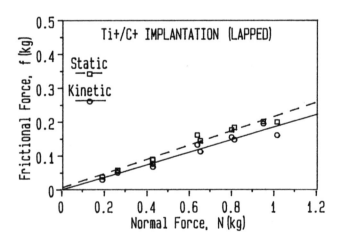

Fig. 6 - Plot of the frictional force (f) versus normal force (N) after the lapped flats of Ti+/C+ implanted S.S. cylinders were pressed against S.S. arch wires (cf Appendix, Fig. vi).

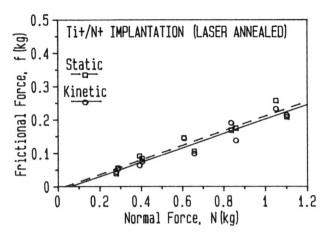

Fig. 7 - Plot of the frictional force (f) versus normal force (N) after the laser annealed flats of Ti+/N+ implanted S.S. cylinders were pressed against S.S. arch wires (cf Appendix, Fig. vii).

respectively) surpassed. Because no AES was performed, interpretation of the results is guarded as uncertainties about the degree of homogeneity and the possibility of having sputtered away some of the implanted ions remain unanswered. Nevertheless, this preliminary experiment shows promise, too.

SUMMARY AND CONCLUSIONS

For the orthodontic patient, malocclused teeth are oftentimes moved by sliding bracketed-teeth along an arch wire. Frictional forces inevitably result that are dependent, in part, on the opposing materials which form the couple. Any improvement in either the efficiency or the reproducibility of force transmission could reduce treatment time and some discomfort associated with wearing "braces." To reduce the detrimental effects of friction, ion implantation was investigated and these conclusions were drawn:

1. Among unimplanted couples, the frictional coefficients (u) increased as stainless steel, cobalt-chromium, nickel titanium, and beta-titanium arch wires were drawn between the flats of stainless steel cylinders.

2. Under the present conditions and with a few exceptions, no improvements in the u´s resulted, when the flats of implanted versus unimplanted stainless steel cylinders were used.

3. Although the effects of saliva on sliding friction require further elaboration, comparisons of data from unimplanted specimens, which have been tested in the dry and wet states, suggest that saliva is not controlling the u´s of implanted specimens.

4. Improvements of the hardness and wear will not necessarily correlate with lower u´s if the normal stresses are light enough so that the u´s are measured on an immediate surface, which is free from carbon contamination.

5. Preliminary results suggest that the removal of the immediate surface layer by lap-

ping, or the laser annealing of the surface layer, may reduce the u´s-- albeit presently not even to the levels of the original unimplanted surfaces.

ACKNOWLEDGEMENT

S.W.A. acknowledges support under NIH grant 2-S07-RR05333. Our thanks to North Carolina State University, Oak Ridge National Laboratory, and Spire Corporation for the ion implantation and to Ormco, Rocky Mountain, and Unitek/3M Corporations for the arch wire materials. Special thanks to Dr. Robert Reeber for his helpful discussions, Dr. Jagdish Narayan for the laser annealing, and Mr. John Whitley and Mr. Warren McCollum for their assistance with the artwork.

REFERENCES

1. Greenberg, A. R. and R. P. Kusy, J. Dent. Res. (Special Issue A) 58, 98 (1979)
2. Sioshansi, P., Mat. Sci. Eng. 90, 373-377 (1987)
3. Sioshansi, P., R. W. Oliver, and F. D. Matthews, J. Vac. Sci. Technol. A, 3 (1985)
4. Hartley, N. E. W., W. E. Swindlehurst, G. Dearnaley, and J. F. Turner, J. Mat. Sci. 3, 900-904 (1973)
5. Singer, I. L. and R. A. Jeffries, Mat. Res. Soc. Symp. Proc. 27, 637-642 (1984)
6. Singer, I. L. and R. A. Jeffries, Mat. Res. Soc. Symp. Proc. 27, 667-672 (1984)
7. Dillich, S. A., R. N. Bolster, and I. L. Singer, Mat. Res. Soc. Symp. Proc. 27, 637-642 (1984)
8. Sioshansi, P. and J. J. Au, Mat. Sci. Eng. 69, 161-166 (1985)
9. Pope, L. E., F. G. Yost, D. M. Follstaedt, S. T. Picraux, and J. A. Knapp, Mat. Res. Soc. Symp. Proc. 27, 661-666 (1984)
10. Follstaedt, D. M., F. G. Yost, and L. E. Pope, Mat. Res. Soc. Symp. Proc. 27, 655-660 (1984)
11. Iwaki, M., Mat. Sci. Eng. 90, 263-271 (1987)
12. Oliver, W. C., R. Hutchings, J. B. Pethica, E. L. Paradis, and A. J. Shuskis, Mat. Res. Soc. Symp. 27, 705-710 (1984)
13. Konishi, R. N., J. Q. Whitley, and R. P. Kusy, Dent. Mater. 1, 55-57 (1985)
14. Kusy, R. P., J. Q. Whitley, M. J. Mayhew, and J. E. Buckthal, Angle Orthod. 58, 33-45 (1988)
15. Kusy, R. P. and J. Q. Whitley, Am. J. Orthod. Dentof. Orthop. (In press)
16. Hutchings, R., W. C. Oliver, and J. B. Pethica, "Surface Engineering", ed. 1, pp. 170-184, Saunders, Inc., New York (1984)
17. Kapila, S., P. Angolkar, M. G. Duncanson Jr., and R. S. Nanda, J. Dent. Res. (Special Issue A) 68, 386 (1989)
18. Kusy, R. P., and J. Q. Whitley, J. Biomech. (In press)
19. Pethica, J. B., R. Hutchings, and W. C. Oliver, Nucl. Instrum. Meth. 209/210, 995-1000 (1983)

20. McHargue, C. J., Int. Met. Rev. 31, 49-76 (1986)
21. Oliver, W. C., Personal communication (1989)
22. Follstaedt, D. M. and S. M. Meyers, Nucl. Instrum. Meth. 209/210, 1023-1031 (1981)
23. Bowden, F. P. and D. Tabor, "Friction and Lubrication", ed. 1, p. 146, John Wiley, New York (1956)
24. Gray, D. E., "American Institute of Physics Handbook", ed. 3, pp. 2-42 - 2-43, McGraw-Hill, New York (1972)
25. Nordling, C. and J. Oosterman, "Physics Handbook", ed. 2, p. 43, Chartwell-Bratt Ltd., London (1982)
26. Pope, L. E., D. M. Follstaedt, J. A. Knapp, and J. C. Barbour, Mat. Res. Soc. Symp. Proc. 100, 175-180 (1988)
27. Deresiewicz, H., In "Approaches to Modelling of Friction and Wear" (F. F. Ling and C. H. T. Pan, eds.), pp. 56-60, Springer-Verlag, New York (1986)
28. Stannard, J. G., J. M. Gau, and M. A. Hanna, Am. J. Orthod. 89, 485-491 (1986)
29. Nicolls, J., Dent. Prac. Dent. Rec. 18, 362-366 (1967/1968)
30. Andreasen, G. F. and F. R. Quevedo, J. Biomech. 3, 151-160 (1970)
31. Reeber, R. R., N. Yu, R. P. Kusy, and W-K. Chu, Appl. Phys. Let. (In press)

APPENDIX

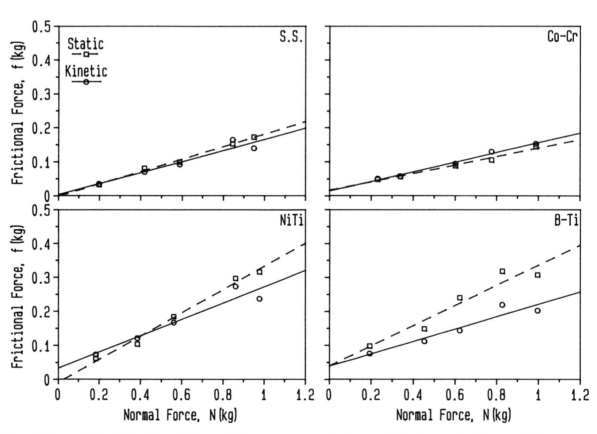

Fig. 1 - Plots of the frictional force (f) versus normal force (N) after the flats of S.S. control cylinders were pressed against each of the four arch wire alloys (cf Table 4).

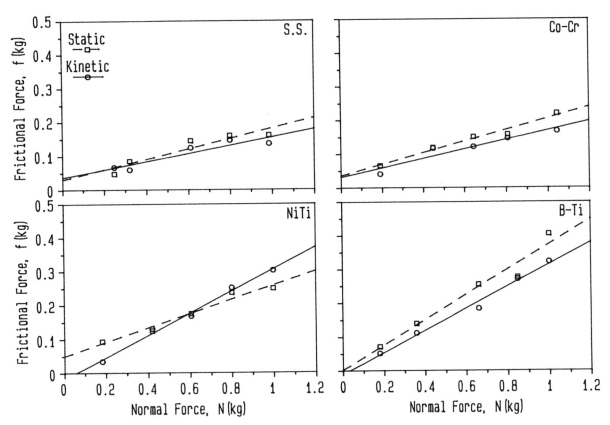

Fig. ii – Plots of the frictional force (f) versus normal force (N) after the flats of N+ implanted S.S. cylinders were pressed against each of the four arch wire alloys (cf Table 4).

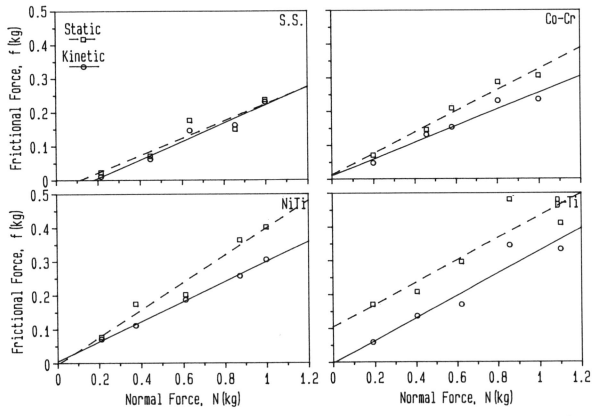

Fig. iii – Plots of the frictional force (f) versus normal force (N) after the flats of N+/Cr+ implanted S.S. cylinders were pressed against each of the four arch wire alloys (cf Table 4).

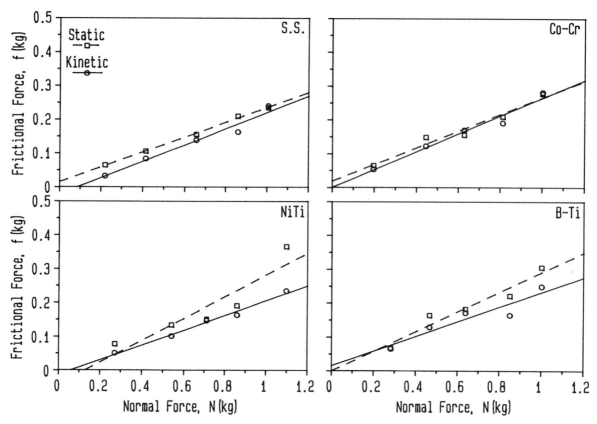

Fig. iv – Plots of the frictional force (f) versus normal force (N) after the flats of N+/C+ implanted S.S. cylinders were pressed against each of the four arch wire alloys (cf Table 4).

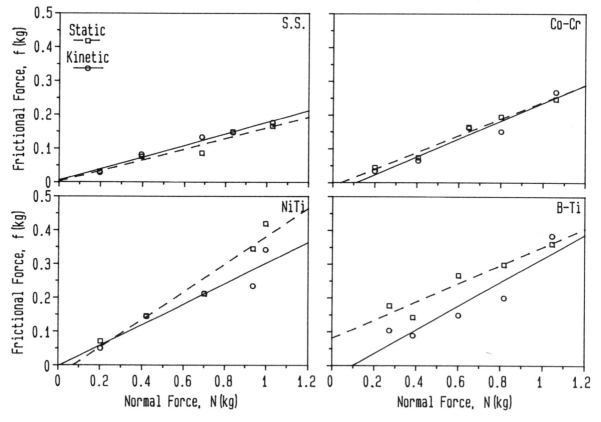

Fig. v – Plots of the frictional force (f) versus normal force (N) after the flats of C+ implanted S.S. cylinders were pressed against each of the four arch wire alloys (cf Table 4).

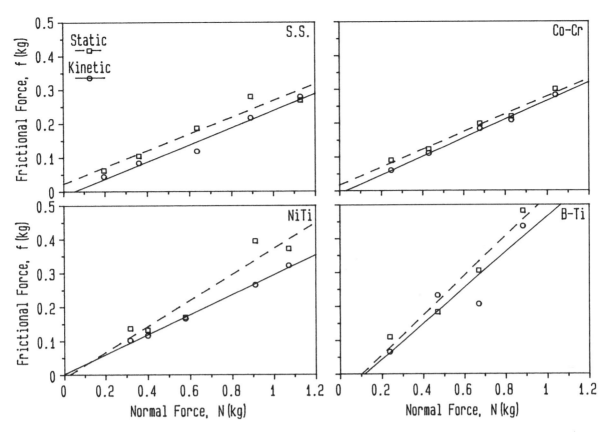

Fig. vi – Plots of the frictional force (f) versus normal force (N) after the flats of Ti+/C+ implanted S.S. cylinders were pressed against each of the four arch wire alloys (cf Table 4).

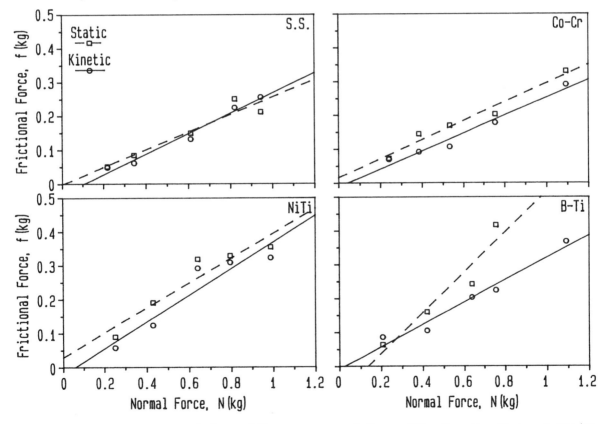

Fig. vii – Plots of the frictional force (f) versus normal force (N) after the flats of Ti+/N+ implanted S.S. cylinders were against each of the four arch wire alloys (cf Table 4).

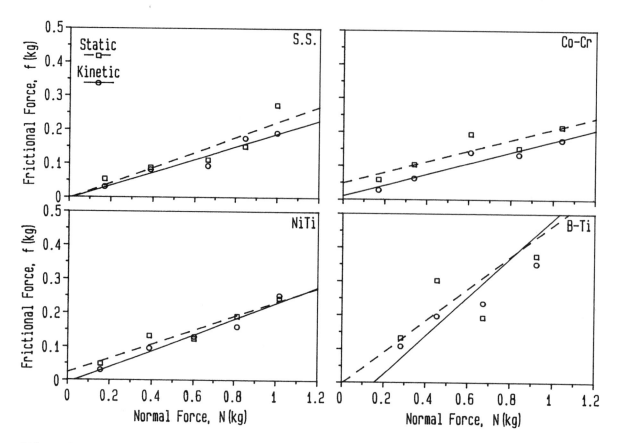

Fig. viii – Plots of the frictional force (f) versus normal force (N) after the flats of the Ti+ implanted S.S. cylinders were pressed against each of the four arch wire alloys (cf Table 4).

ION NITRIDING OF TITANIUM AND Ti-6Al-4V ALLOY

K. T. Kembaiyan, R. D. Doherty
Department of Materials Engineering
Drexel University
Philadelphia, Pennsylvania, 19104, USA

R. P. Singh
Wyman and Gordon Company
North Grafton, Massachusetts, USA

R. Verma
GloTech Inc.
Newton, Pennsylvania, USA

ABSTRACT

Pure titanium and Ti-6Al-4V alloy were ion nitrided under pure nitrogen plasma. The effect of various process parameters i.e., temperature, gas pressure, time and voltage on the thickness of the nitrided layers have been studied. Growth kinetics of the nitrided layers were studied by monitoring the changes in thickness and microhardness. On both titanium and Ti-6Al-4V, multilayer coatings were found. These consisted, as expected from the Ti-N phase diagram, of TiN, Ti_2N and α case consisting of high concentration of nitrogen in the hcp phase. All these layers are very hard with the hardness falling from 1800 Vickers Knoop in the TiN layer on Ti-6Al-4V, through 900 in the α case to 350 in the unmodified case. Strong metallurgical bonding appears to occur between various layers. The growth kinetics show a $t^{1/2}$ time dependence characteristics of control by nitrogen diffusion in the outer layers. Wear tests showed reduced weight loss by 30 to 50 times of the nitrided materials compared to the unnitrided samples. Highly wear resistant coating of titanium and Ti-6Al-4V are therefore readily achieved by plasma ion nitriding.

SEVERAL PROCESSES ARE CURRENTLY USED to provide wear protection to metal surfaces in contact, and to extend the life of critical components used in bearings, cutting tools and aerospace industry (1). Hard coatings, which can withstand high stress without plastic deformation and fracture, are more effective in reducing friction and wear than soft layers. Despite their attractive engineering properties, titanium and Ti-6Al-4V alloys appeared to lack good wear resistance. Formation of a nitride coating has been achieved by means of ion nitriding (2-7) ion implantation (8-11) and gas nitriding (12-16) which may eliminate this problem by providing a strong and adherent coating. Among various processes, plasma ion nitriding is one of the most successful process in which a adherent nitride layer on the surface of the part is obtained by subjecting the part to a nitrogen plasma at a sufficient temperature for nitrogen diffusion. Plasma nitriding has the advantage of providing close dimensional tolerance and low cost. Metin and Inal (5) have proved that ion nitriding technique could be successfully used for titanium and Ti-6Al-2Sn-4Zr-2Mo alloys. The present investigation was aimed at producing stable nitride coatings on commercially pure titanium and on the Ti-6Al-4V alloy.

MATERIALS AND METHODS

Chemical compositions of Ti-6Al-4V alloy is given in Table 1. Most of the samples used were of 13mm in diameter and 13mm in thickness. Blocks used for wear tests were of 16 mm x8 mm x6 mm in dimensions. The specimens were polished using 600 grit silicon carbide paper and

Table 1. Chemical composition of Ti-6Al-4V alloy used in this study

Fe(wt.%)	C(wt.%)	N(wt.%)	O(wt.%)	H(wt%)	Al(wt.%)	V(wt.%)
0.20	0.02	0.018	0.19	0.076	6.2	3.9

cleaned in acetone and alcohol prior to nitriding.

A schematic diagram of the reactor assembly used in this study is shown in Fig. 1. This unit consists of three main parts, an external heating unit to maintain a constant temperature, an external power supply to strike and maintain a plasma and a vacuum gas system to regulate the partial pressure of nitrogen and argon. The d.c. voltage necessary to obtain good nitriding glow is between 500 and 730 V. The furnace is maintained at a (+)ve potential and the work pieces to be nitrided is placed on the cathode. The nitrogen gas pressure was maintained at 3 torr. The nitriding temperature was measured using a chromel-alumel themocouple embeded about 10mm into the stainless steel cathode plate very close to the sample location and controlled by changing the power input. Nitriding was carried out in the temperature range between 700 °C and 900 °C as measured by the thermocouple. Energy for substantial diffusion of nitrogen into the work piece is provided by both the kinetic energy transfer generated by the ion bombardment and externally supplied heat. The actual temperature of the sample may therefore be slightly higher than the furnace temperature due to the plasma heating.

The cross-sectional samples of nitrided pieces were mounted in baklite/SiC mixture to ensure edge retention. Specimens were ground through successive SiC papers, grades 240, 320, 400, and finally polished with a 6 micron diamond paste and finished with a 1 micron diamond cloth. Polishing times were limited to 30 - 45 seconds on each wheel in order to avoid the onset of unacceptable bevelling of the specimen edges. Rotating grinding wheels were found to be too severe for the thin nitrided layers, resulting in heavy damage and fracture of the layers, and gentle hand grinding on SiC strips, without rotation of specimen between paper grades, was found to be effective. The nitrided specimens were etched in a mixture of 2 wt% HNO_3 and 2 wt% HF. Thickness of various layers formed during nitriding was measured using photomicrographs taken at different magnifications using standard optical microscopes and a scanning electron microscope. X-ray diffractometry was performed for phase identification. Microanalysis across the cross section of the nitrided layers was conducted using a wavelength dispersive spectrometer and Auger electron spectroscopy. The microhardness indentations were made by means of a Leco M-400 microhardness tester. A 200g load was used to provide optimum indentation depths across various layers.

Adhesive wear and friction coefficients were measured using a block-on-ring tester (17). The test conditions are detailed in Table 2. The test block is loaded against a ring which rotates at a given speed. The block and the ring were cleaned in trichloroethane and methanol using an ultrasonic cleaner before and after the test, dried and weighed by means of an analytical balance to the nearest 0.1 mg. The scar width on the block is measured using an optical microscope. The friction force required to keep the block in place is contineously measured during the test with a load cell. The wear rate was evaluated by the weight loss of both the block and the ring. The coefficient of friction (μ) was calculated from the applied contact force and resulting drag force.

RESULTS AND DISCUSSION

CONDITIONS OF ION NITRIDING - The high affinity of titanium for oxygen and carbon necessiates clean equipment and experimental techniques. All the parts, with the exception of the molybdnum heating elements in the interior of the reactor chamber shown in Fig. 1 were made of stainless steel. The source and the carrier gases used, ultrahigh purity grade nitrogen and argon were passed through 'gas purifiers' prior to entering into the reactor chamber. The use of pure titanium 'getter' turnings further minimized the residual oxygen. Prior to heating the furnace, it was purged with argon and nitrogen. The work pieces were being sputter cleaned as the chamber approached the specified nitriding temperature. The amount of argon was gradually decreased with increasing the flow of nitrogen while maintaining the total pressure close to about 3 torr in the chamber. Following these procedures, small parts of pure titanium and Ti-6Al-4V were successfully nitrided giving an attractive golden color. Incorporation of oxygen in the system develops a white oxide band on the surface and carbon contaminates with a unsightly black surface. A striking result was the successful nitriding of all surfaces of the titanium samples including the lower surface in contact with the stainless steel cathode. This result indicates either plasma penetration between the sample and the cathode or more likely rapid transport by surface diffusion of nitrogen atoms along the nitrided surfaces. A drawback is that this phenomenon has so far prevented the development of any suitable masking technique. It might be noted that titanium at the nitriding temperature diffusionally bonds to most possible masking agents.

MICROSTRUCTURAL FEATURES - Plasma nitrided Ti-6Al-4V at 900 °C for 16 hours resulted in the formation of several distinct phases on the surface as shown in the optical micrograph of the cross-sectional sample in Fig. 2. Nitride layers grew parallel to the surface of the work pieces. The overlayers consisted of TiN phase followed by a thin layer of Ti_2N and subsequently a thick layer of solid solution of nitrogen in titanium in both pure titanium and Ti-6Al-4V as evidenced by the X-ray diffraction analysis. Formation of these nitride layers is consistant with the phase diagram of Ti-N (18). The dual phase $(\alpha+\beta)$ microstructure of the alloy is retained in the core as shown in Fig. 2b.The growth of various nitride layers in pure Ti and

Ti-6Al-4V under various nitriding conditions are illustrated in Figs.3 through 7. The optical micrographs of the cross-sections of pure titanium in Fig. 3 shows a gradual increase in the thickness of various layers with increasing nitriding temperature for 16 hours of nitriding. Micrographs of Ti-6Al-4V in Fig. 4 exhibits similar microstructural features with slightly thinner nitrided layers. The intermediate Ti_2N phase is more prominent in the alloy than in the pure titanium.

GROWTH KINETICS - Kinetics of nitride layer growth on pure titanium and Ti-6Al-4V was studied at above and below the β transformation temperature, 882 °C in pure titanium. The overall growth of all the distinct layers (TiN, Ti_2N and the α-stabilized layer) obeyed a simple $t^{1/2}$ relationship as shown in Fig. 5 for pure titanium and in Fig. 6 for Ti-6Al-4V. These results indicate that the growth kinetics is diffusion controlled. The growth rates of compound layer (TiN + Ti_2N) in titanium and Ti-6Al-4V at 900 °C are compared in Fig. 7, which reveals again that the growth rate is controlled by a diffusional transport through the product layers.

Microanalysis along the cross section using wavelength dispersive spectrometry (WDS) revealed the depletion of vanadium in the nitrided layer in Ti-6Al-4V. The partitioning of vanadium across the nitrided layers is further substantiated by the Auger electron spectroscopy (AES) as shown in Fig. 8. The significant vanadium peak present in Fig. 8a in the core can no longer be seen in the nitride layers in Fig. 8b. In this dual $(\alpha + \beta)$ phase alloy the aluminum stabilizes the α phase while vanadium acts as a β stabilizer. Thus the depletion of vanadium in the vicinity of the surface stabilizes the hcp α phase and nitriding results in mostly a solid solution of nitrogen in the α phase. Precipitation of AlN in the nitrogen rich case is possible but so far this has not been seen by X-ray diffraction. TEM studies of the nitrided layers will be required to test this possibility.

HARDNESS MEASUREMENTS - To evaluate the effects of nitriding temperature and time on the extent of nitrogen dissolution, microhardness traverses were carried out on samples which has been heat treated at 700, 800 and 900 °C for different exposure times. The effect of nitriding temperature on the microhardness of pure titanium and Ti-6Al-4V is illustrated in Figs. 9 and 10. A significant increase in the Knoop hardness upto 1800 is observed at the surface when compared with about 350 at the core in Ti-6Al-4V with increasing temperature. Nitriding at higher temperature also resulted in a higher hardness which may be due merely to the increased thickness of the hardened layers. The gradual decrease in the hardness toward the core indicates the decreasing amount of nitrogen in the nitride layers. All of the contours tend to be steep near the interface, indicating increasing concentration of nitrogen in such regions. Similar investigations involving hardness measurements across the nitrided layers enabled earlier studies (5,12) to calculate the nitrogen diffusion profiles and diffusion coefficients of nitrogen in the various layers. The results in the present work are in agreement with previous studies (5,12) which indicated that the rate controlling mechanism in the growth of nitride layers is diffusion of nitrogen through the product layers.

NITRIDING OF SHEETS - Following the progress of bulk samples, thin sheets or fiber/wires of titanium and Ti-6Al-4V were ion nitrided. These starting materials either can be partially nitrided or fully nitrided by controlling the operating parameters. The partially nitrided structures are themselves composite structures. However, these nitrided thin sheets, wires, fibers and filaments can also be used as starting materials for fabricating of higher level composite structures. Fig. 11(a) shows an optical micrograph of a nitrided Ti-6Al-4V

sheet. Depending on the sheet thickness and process parameters, the layer thickness and the nitrogen content in the core can be modified to suit to the particular applicability. It is found that nitriding very thin sheets causes it to buckle after nitriding. Details of the responsible cause are not known at present but it is is likely to be due to residual stresses developed by the nitriding process. The growth of nitrided layer on the ribbon was faster than a bulk sample due to rapid saturation of the ribbon by nitrogen. It should be recognized that the high solubility of nitrogen in titanium will mean that nitride layers on bulk samples of titanium will be unstable at high temperatures. In nitrogen saturated thin samples the surface layers will, however, be stable.

Nitrided sheets can be sandwiched with other materials such as Ti, Ti-6Al-4V, Al and Al based alloys depending on the final application. Fig. 11(b) shows a nitrided Ti-6Al-4V sheet used as a composite in an aluminum matrix formed by conventional vacuum hot pressing.

WEAR TEST - Although work has been carried out on the wear resistance of nitrided steels there is very little or no data published on the wear resistance of ion nitrided Ti and Ti-6Al-4V alloy. Martinella et al.(8) have compared the wear behavior of nitrogen ion-implanted and gas nitrided Ti-6Al-4V alloy. It was observed that rough nitrided surfaces showed better wear resistance than lapped nitrided surfaces or lapped implanted surfaces. The authors attributed this improvement in the wear resistance to the reduction in friction induced by chemical modification of the surface as a result of oxide and TiN. In the present investigation, a block-on-ring adhesive wear test was used to evaluate the

Table 2. Block-on-ring wear test parameters

Block	: Nitrided and unnitrided samples of Ti and Ti-6Al-4V
Ring	: Uncoated steel (4620)
Speed	: 180 rpm
Load Static	: 30 lbs to 180
Dynamic	: 6 lbs
Temperature	: 23 °C
Test Environment	: Unlubricated, air at relative humidity 85%

wear resistance of the nitrided samples. This is one of the most widely used industrial techniques for evaluating the adhesive wear resistance of components. The weight loss of the unnitrided and plasma nitrided wear blocks is compared in Fig. 12 which indicates that <u>the plasma nitride block is an order of magnitude or more superior than the unnitrided block</u>. A slight reduction of the coefficient of dynamic friction was also observed in the nitrided titanium (0.14) and Ti-6Al-4V alloy (0.15) when compared with unnitrided samples (0.23 and 0.20). Although a brief nitriding gives a major benefit, the results show that with longer time and thicker nitrided layer, wear resistance can be further improved. The scar widths of the unnitrided and ion nitrided blocks of titanium and Ti-6Al-4V are compared in Fig. 13 which illustrates the major improvement in the wear resistance of the nitrided samples. In the nitrided blocks, major wear groves are seen while after nitriding only a very narrow scar is seen.

CONCLUSIONS

1. Pure titanium and Ti-6Al-4V alloy were successfully ion nitrided and the process parameters established for small components. An attractive gold color on the samples was produced by this process.
2. Ion nitriding offers a hard case of as high as 1800 Knoop hardness as compared with 350 of the core. The interlayers have intermediate hardnesses.
3. Various nitrided layers, TiN, Ti_2N and a solid solution of nitrogen in the α-phase in sequence from the surface were identified by X-ray diffraction
4. Optical microscopy and microhardness measurements indicate that the growth rate of nitrided layers is controlled essentially by the diffusion of nitrogen through the nitrided layers.
5. Wear tests indicated that the nitrided parts exhibited excellent wear resistance when compared with unnitrided part. 30 to 50 times less weight loss was found for the nitrided samples.
6. Ion nitriding therefore allows the addition of wear resistance to the other attractive engineering properties of titanium and its alloys.

ACKNOWLEDGEMENT

This work is funded by Advanced Technology Center, Philadelphia and GloTech Inc. Newtown, Pa. under Ben Franklin Partnership Program. Our thanks are due to Mr. Tim Vela and J. Panchal of Alloy Technology for their support in carrying out the wear tests and M. Dipetro for performing some of the experiments.

REFERENCES

1. H.E. Hitermann, J. Vac. Sci. Technol. B2 816 (1984).
2. K.T. Rie and Rn. Lampe, Mat. Sci. & Eng., 69, p.473 (1985).
3. M. Konuma, O. Matsumoto, J. Less Common Metals, 52 p. 145 (1977)
4. E. Rolinski, Surf. Eng., 2 (1), p.35 (1986).
5. E. Metin, O.T. Inal "Ion nitriding " Conf. Proc. ASM Int., T. Spalvins (Ed.), p.61 (1986).
6. C. Badini, D. Mazza and T. Bacci, Mat. Chem. & Phy. 20, p.559 (1988).
7. R. Avni and T. Spalvins, Mat. Sci. & Eng., 95, p.237 (1987).
8. R. Martinella, S. Giovanardi, G. Chevalland, M.Villani, A. Molinar and C. Tosella, Mat. Sci. & Eng., 69, p.247 (1985).
9. K. Hohmuth and B. Rauschenbach, Mat. Sci. & Eng., 69, p.489 (1985).
10. R.G. Vardivan and R.A. Kant, J. Appl. Phys., 53 (1), p.690 (1982).
11. R. Hutchings, Mat. Sci. & Eng., 69, p. 129 (1985)
12. N.G. Baishiua, D.P. Shashkov, E.M. Karna and Y.M. Mikhalin, Met. Sci. & Heat Treat. 23 (7-8), p.503 (1981).
13. J.L. Wyatt and N.J. Grant, "Source book of Nitriding", ASM, p.191 (1977).
14. K.N. Strafford and J.M. Towell, Oxidation of Metals, 10, (1), p.41 (1976).
15. E. Rolinski, J. Less Common Metals, 141, L11 (1988).
16. J.P. Bars, D.David, E. Elchessahar and J. Debuigne, Met. Trans. 14A, p.1537 (1983).
17. ASTM G77-83.
18. "Titanium Alloys", (Ed). E.W. Collings, ASM, Metals Park, p.64 (1984).

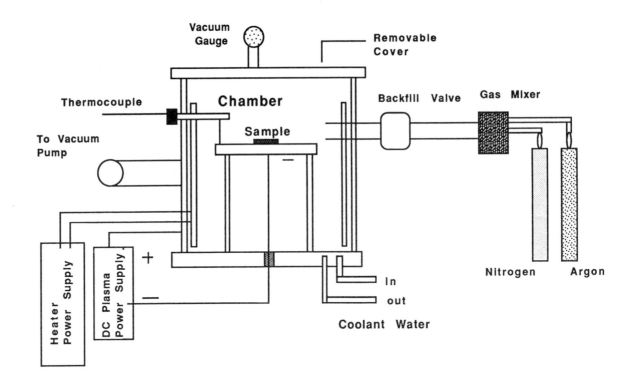

Fig. 1. Schematic diagram of the plasma nitriding apparatus.

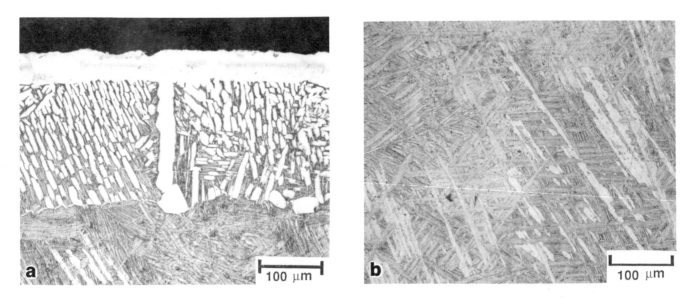

Fig. 2. Optical micrographs of the cross sections of Ti-6Al-4V alloy ion nitrided at 900 °C for 16 hours. (a) case region. (b) core region.

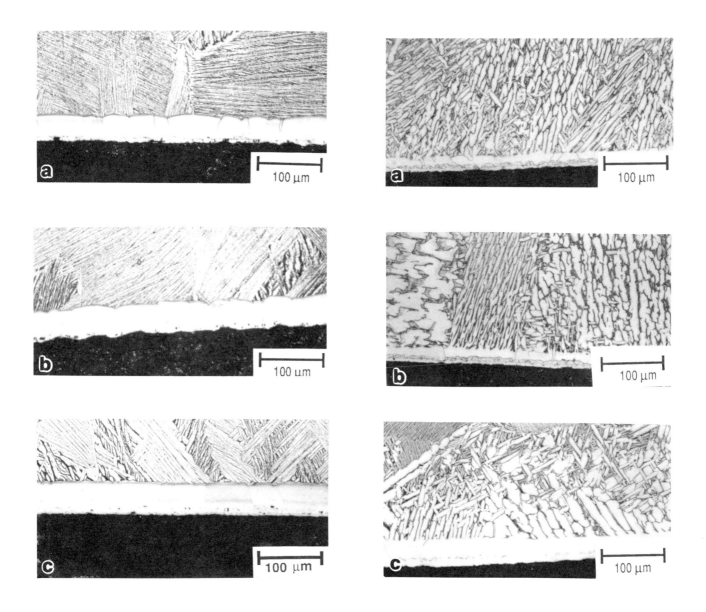

Fig. 3. Cross section of commercial purity titanium ion nitrided for 16 hours. (a) 700 °C. (b) 800 °C. (c) 900 °C.

Fig. 4. Cross sections of Ti-6Al-4V alloy ion nitrided for 16 hours. (a) 700 °C. (b) 800 °C. (c) 900 °C.

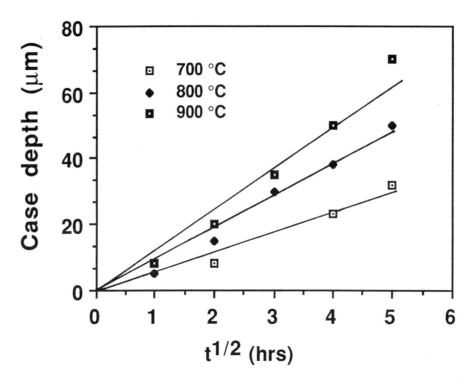

Fig. 5. Case depth of pure titanium as a function of square root of time.

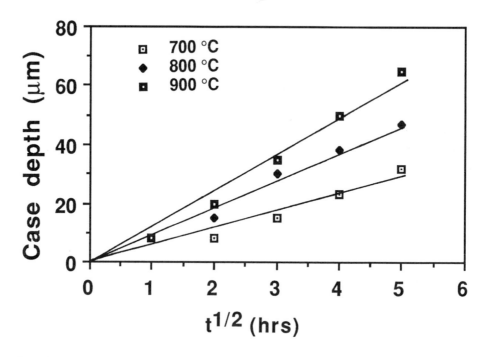

Fig. 6. Case depth of Ti-6Al-4V sample as a function of square root of time.

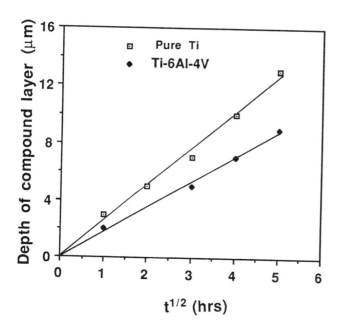

Fig. 7. The depth of the compound layers of pure titanium and Ti-6Al-4V alloy nitrided at 900 °C as a function of time.

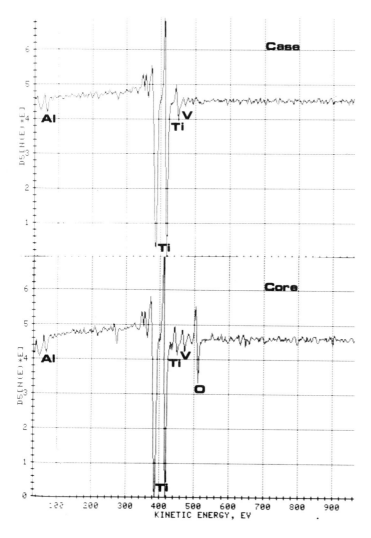

Fig. 8. AES spectra obtained from the case and core regions of a Ti-6Al-4V sample nitrided at 900 °C for 16 hours.

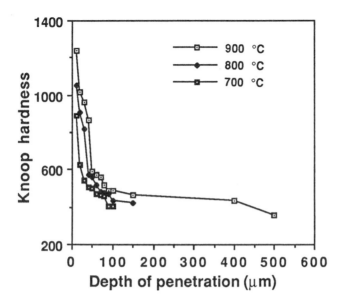

Fig. 9. Variation of the microhardness across the cross section of a pure titanium sample nitrided for 16 hours at specified temperatures.

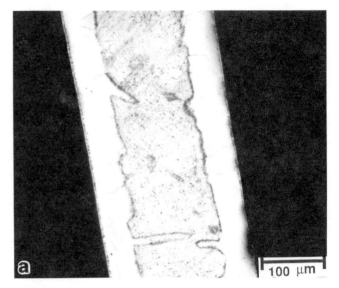

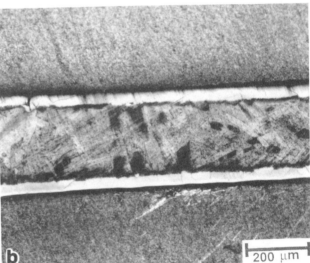

Fig. 11. (a) Cross section of a nitrided ribbon of Ti-6Al-4V alloy (900 °C, 9 hrs). (b) Nitrided ribbon of Ti-6Al-4V used as a composite in an aluminum matrix.

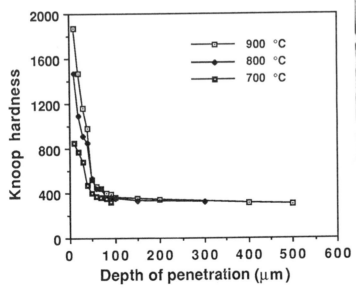

Fig. 10. Variation of microhardness across the cross section of a Ti-6Al-4V sample nitrided for 16 hours at specified temperatures.

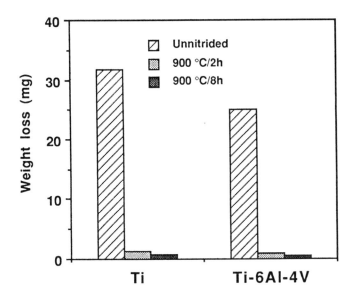

Fig. 12. Comparison of weight loss of nitrided and unnitrided samples of pure titanium and Ti-6Al-4V alloy in block-on-ring tests.

a

b

Fig. 13. Unnitrided and ion nitrided blocks of (a) pure titanium and (b) Ti-6Al-4V alloy subjected to wear tests showing worn groves in the unnitrided samples and the surface scar widths in the nitrided samples.

PLASMA SINTERING OF EXPLOSIVE SHOCK ACTIVATED Al$_2$O$_3$-ZrO$_2$ POWDER, Al$_2$O$_3$ POWDER-SiC WHISKER, AND Al$_2$O$_3$/ZrO$_2$ POWDER-SiC WHISKER MIXTURES

Murat Bengisu, Osman T. Inal
Materials & Metallurgical Engineering Department
New Mexico Institute of Mining & Technology
Socorro, New Mexico, 87801, USA

Abstract

Plasma sintering of shock treated and unshocked Al$_2$O$_3$–based ceramic composites was conducted for various durations in several plasma gases. Al$_2$O$_3$ –10%ZrO$_2$ composites were sintered to above 90% fractional density in 3 to 5 minutes. The highest fractional density achieved in Al$_2$O$_3$–30SiC$_w$ composites was 64%, obtained by shock treatment of the powders at 10 GPa prior to plasma sintering. Increasing the sintering time of whisker containing composites did not increase the amount of consolidation. The most suitable gases for HCD–plasma sintering were H$_2$ and air among four types of gases in this study. Microstructures of plasma sintered Al$_2$O$_3$–10%ZrO$_2$ composites comprised of fine equiaxed grains with low porosity. Outer regions of unshocked / plasma sintered Al$_2$O$_3$–30%SiC$_w$ composites revealed a high density and uniform microstructure characteristic of successful sintering.

PLASMA PROCESSING OF MATERIALS is a relatively new area where significant research is being currently made. Some plasma technologies find widespread applications due to the unique material properties achieved with them and the simplicity of the processing. Besides the better known ion nitriding [1-3]/carburizing [4] and plasma assisted deposition techniques [5], there are numerous processes which make use of glow discharge exposure . Sputtering, activated reactive evaporation, plasma etching, plasma polymerization [5], and synthesis of ceramic powders [6,7], are some of these processes.

Another potential use of the glow discharge is sintering of ceramic materials. Three different plasma methods have been used for sintering of various ceramics. In historical order, these are microwave induced plasma (MIP) [8], hollow–cathode discharge (HCD) [9], and induction coupled plasma (ICP) [10] sintering. The significance of plasma sintering by any of these processes lies in the unusual rapid sintering rates achieved, uncommon to conventional densification methods. Rapid sintering of various ceramics including Al$_2$O$_3$ [8-14], Y$_2$O$_3$-PSZ [14], BeO$_2$ [8], HfO$_2$ [8], UO$_2$ [15,16], and ZrO$_2$ [17], has been demonstrated by a number of researchers. Uniform, fine microstructure and better mechanical properties are commonly observed in plasma sintered ceramics with fractional densities ranging from 96 to 100% [9-14].

The present study utilizes HCD–plasma sintering to compare sintering kinetics of shock treated and unshocked Al$_2$O$_3$ based composites. Shock treatment of ceramic powders is known to enhance their sinterability by the introduction of defects and/or strain to the structure [18] as well as by particle communition [19]. In this study, the densification kinetics during plasma sintering of shock treated and unshocked powders was expected to provide new information on both the sintering process and the role of activation in the sintering of ceramics. In addition, the feasibility of consolidating whisker reinforced ceramics by the plasma sintering process was investigated. Mechanical properties of resultant compacts were evaluated and correlation to process parameters was attempted.

EXPERIMENTAL PROCEDURE

High purity ZrO_2 * (no stabilizer added) and/or SiC whiskers ** were mixed with α-Al_2O_3 powders *** by ball milling for 12 hours in pH-controlled distilled water solution. The slurries were stir-dried on a hot plate. Dry mixtures were deaggregated using a mortar and a pestal.

Powder mixtures were pressed into cylindrical steel containers to about 50 % fractional density and shock treated at 4 and 10 GPa. Axisymmetric shock compression was applied utilizing suitable amounts of ANFO explosive. Reported shock pressures are calculated values and represent pressures generated at the container walls. Shock compressed steel containers were sectioned via high speed abrasive SiC discs. Recovered powders were remilled until no visible agglomerate remained.

Four to five mm diameter and 25 mm long rods were prepared by plastic forming of pastes made out of shock treated and unshocked powder-alcohol mixtures. The experimental set up used for HCD-plasma sintering studies is shown in Figure 1. A mechanical vacuum pump was used to evacuate the bell

---

* –SF-ULTRA ZrO_2, Z-TECH Corporation, Bow NH

** – ARCO SiC whiskers

*** – RC-HP-DBM, Malakoff Ind.,Inc., West Malakoff, TX

jar to a pressure of 0.2 Torr. Except for sintering experiments done in air, the system was flushed twice with the plasma gas. A power supply with a capacity of 2 kW and 1 A output was used to generate the plasma. The application of DC voltage to the system generates the plasma. Electron bombardment of the sample placed at the center of two concentric cylinders, that comprise the anode and cathode, is assumed to be the main source of heat that affords the sintering [9].

Temperature of the samples was measured by a thermocouple inserted into the samples and isolated from the plasma by a protective ceramic rod. An optical pyrometer was also used to measure the surface temperature. Pyrometer measurements were calibrated to account for plasma and bell jar absorption. Application of high voltage for longer than 5 minutes increased the possibility of arcing. After holding the samples at sintering temperatures for desired durations and cooling down to 500°C, the system was opened to air by a bleeding valve.

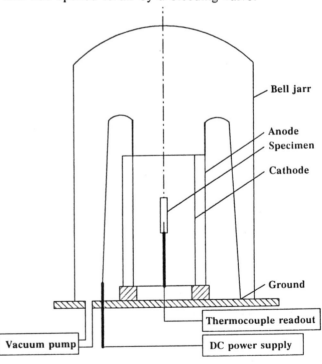

Fig.1. Experimental arrangement used for HCD-plasma sintering studies

Volumes of sintered samples were evaluated by differential weighing of wet samples in water and air after boiling for 4 hours and cooling to room temperature [20]. Fractional densities were calculated from dry weights and measured volumes of the specimens. Samples were cut by diamond abrasive blades, mounted in epoxy, and successively polished using different grades of diamond paste. Mechanical properties were determined by Knoop and Vickers indentations. Fracture toughness values were measured from microcracks developed from indentations produced by a Vickers indentor [21,22] at 12 kg load. Elastic modulus values were determined from Knoop indentations [23].

RESULTS

DENSIFICATION KINETICS – Sintering of Al_2O_3–$10ZrO_2$ composites was accomplished in less than 3 minutes at about 1300°C. The maximum fractional density achieved was 94.3% for unshocked Al_2O_3–$10ZrO_2$ composite and 98% in dense regions (agglomerates) of 10 GPa shock treated Al_2O_3–$10ZrO_2$ composite. A similarly preconsolidation processed, unshocked Al_2O_3–$10ZrO_2$ composite had a fractional density of 94.4% after conventional sintering at 1450°C for 12 hours.

Four plasma gases were used to sinter unshocked Al_2O_3–$10ZrO_2$ powders, namely air, O_2, H_2, and Ar. 4 GPa shock treated powders were sintered in O_2, H_2, and air plasmas. Fractional densities achieved with different plasma supporting gases are given in table I. H_2 and air plasmas seem to be more effective in sintering compared to O_2 and Ar, although in unshocked Al_2O_3–$10ZrO_2$ composites the effect is not significant. Sample heating achieved by either gas was not much different from one to another.

The effect of sintering time on the amount of densification was analyzed both in Al_2O_3–$10ZrO_2$

and Al_2O_3–$30SiC_w$ mixtures. Increasing the hold time at the sintering temperature did not increase the overall fractional density of either composite (Tables I and II). Nevertheless, the Vickers hardness of Al_2O_3–$10ZrO_2$ increased from 10.8 GPa to 14.3 GPa and the elastic modulus increased from 204 GPa to 243 GPa, which may be an indication of a better bond of the particles, i.e. better sintering. The density of the Al_2O_3–$30SiC_w$ composite decreased slightly with extended exposure to the plasma. A rather sharp decrease in hardness was measured in samples sintered for longer durations than 5 minutes. These observations suggest that prolonged exposures to the plasma can be detrimental to the sintering of Al_2O_3–$30SiC_w$ composites. This may be attributed to evaporation or deterioration of SiC whiskers during electron bombardment in the DC plasma.

SiC_w containing composites did not sinter to over 65% of their theoretical density. An identical maximum fractional density was achieved in conventionally sintered Al_2O_3–$30SiC_w$ composite. Both plasma and conventionally sintered composites having the highest final densities were prepared from powder

Table I. Properties of shock treated and unshocked plasma sintered Al_2O_3–$10\%ZrO_2$ composites

Shock Pressure (GPa)	Plasma Gas / time (min)	% Theoretical Density	HV * (GPa)	HK ** (GPa)	E † (GPa)	K_{Ic} ‡ (MPa√m)
–	Ar / 5	89.6	12.8	11.9	169	2.2
	H_2 / 5	93.4	11.6	9.8	168	2.1
	O_2 / 5	90.1	11.3	11.2	204	2.3
	air / 3	90.3	10.8	11.1	205	2.2
	air / 5	92.1	13.2	7.9	70	1.4
	air / 10	81.0	14.3	12.4	244	2.5
4	H_2 / 5	82.3	13.9	11.0	211	1.9
	O_2 / 5	71.1	6.5	11.0	236	–
	air / 5	79.8	10.9	9.8	156	1.6
10	H_2 / 5	75.1	–	12.8	166	–

* – 5 measurements for each determination, S.D. ±10% of average
** – 10 measurements for each determination, S.D. ±20% of average
† – 10 measurements for each determination, S.D. ±50% of average
‡ – 5 measurements for each determination, S.D. ±10% of average

Table II. Properties of shock treated and unshocked plasma sintered Al$_2$O$_3$-30%SiC$_w$ and Al$_2$O$_3$-10%ZrO$_2$-30%SiC$_w$ composites

Composite Type	Shock Pressure (GPa)	Plasma Gas / time (min)	% Theoretical Density	HV * (GPa)
Al$_2$O$_3$-30%SiC$_w$	–	H$_2$ / 5	59.0	1.2
	–	H$_2$ / 10	58.9	0.8
	–	H$_2$ / 15	57.7	0.7
	4	H$_2$ / 5	60.5	0.4
	10	H$_2$ / 5	63.6	0.9
Al$_2$O$_3$-10%ZrO$_2$ -30%SiC$_w$	–	H$_2$ / 5	38.4	0.4
	4	H$_2$ / 5	55.0	1.1

* – 5 measurements for each determination, S.D. $\overline{+}$10% of average

Fig.3. SEM micrograph of 10 GPa shock treated plasma sintered Al$_2$O$_3$-30%SiC$_w$ composite (5 minutes in H$_2$ plasma)

mixtures shock treated at 10 GPa. Microstructural examination of an unshocked, plasma sintered Al$_2$O$_3$-30SiC$_w$ revealed that although the core of the composite remained highly porous, a very high density (>95% fractional density) skin was formed by the plasma sintering process (Fig.2). The microstructure of shock treated and plasma sintered Al$_2$O$_3$-30SiC$_w$, however, did not reveal such a high density skin but rather many well sintered powder and powder-whisker agglomerates (Fig.3).

Fig.2. SEM micrograph of outer region of plasma sintered unshocked Al$_2$O$_3$-30%SiC$_w$ composite (5 minutes in H$_2$ plasma)

The effect of shock treatment on the densification rates of all three Al$_2$O$_3$-based composites were studied by the plasma sintering method. Fractional densities of sintered Al$_2$O$_3$-10ZrO$_2$ composites decreased as the shock pressure increased (Table I). In contrast, higher fractional densities were achieved as the shock pressure was increased in SiC$_w$ containing composites. These results are in good agreement with comparative conventional sintering studies for all composite types, except shock treated Al$_2$O$_3$-10ZrO$_2$ composites, which yielded higher densities than their unshocked counterparts (Table III).

MICROSTRUCTURE – Fracture surfaces of Al$_2$O$_3$-based composites were studied by scanning electron microscopy both for fracture and microstructure analysis. The microstructure of unshocked, plasma sintered Al$_2$O$_3$-10ZrO$_2$ comprised of fine, faceted grains and a few small (1.5 μm diameter) pores at triple or quadruple contact points. A careful analysis of the fracture surface revealed several jagged grains indicating transgranular fracture, whereas the major-

Fig.4. SEM micrograph of unshocked plasma sintered Al_2O_3–10%ZrO_2 composite (5 minutes in H_2 plasma)

Table III. Comparison of properties achieved by conventional sintering and by plasma sintering of shock treated and unshocked alumina matrix composites

Composite Type	Shock Pressure (GPa)	Sintering Method	% Theo. Density	HV* (GPa)	HK** (GPa)	E† (GPa)	K_{Ic}‡ (MPa√m)
Al_2O_3 –10% ZrO_2	–	conv. sint. 1450°C/ 12 hrs. in N_2	94.4	18.0	14.3	282	3.0
	4		96.4	18.3	16.2	303	3.0
	10		91.6	8.7	6.8	84	2.5
	–	plasma sint. 5 min. in H_2	93.4	11.6	9.8	168	2.1
	4		82.3	13.9	11.0	211	1.9
	10		75.1	–	12.8	166	–
Al_2O_3 –30% SiC_w	–	conv. sint. 1450°C/ 12 hrs. in N_2	55.6	0.6			
	4		53.4	0.7			
	10		63.8	0.9			
	–	plasma sint. 5 min. in H_2 hrs. in N_2	59.0	1.2			
	4		60.5	0.4			
	10		63.6	0.9			
Al_2O_3 –10% ZrO_2 –30% SiC_w	–	conv. sint. 1450°C/ 12	44.9	–			
	4		56.6	–			
	–	plasma sint. 5 min. in H_2	38.4	0.4			
	4		55.0	1.1			

   * – 5 measurements for each determination, S.D. ±10% of average
  ** – 10 measurements for each determination, S.D. ±20% of average
  † – 10 measurements for each determination, S.D. ±50% of average
  ‡ – 5 measurements for each determination, S.D. ±10% of average

ity of the grains were separated intergranularly (Fig.4).

Shock treated Al_2O_3–10ZrO_2 composites yielded well sintered areas with large pores (several hundred microns) in between. Inhomogeneous sintering was more pronounced in 10 GPa shock treated composites. This type of microstructure is common in many shock treated ceramics and attributed to agglomerates formed by the shock treatment [18,24].

Individual grains of 4 GPa shock treated and plasma sintered Al_2O_3–10ZrO_2 composite had higher curvature than those in the unshocked composite (Fig.5). The fracture surface was much smoother and no transgranular fracture was observed, indicating that complete bonding of the particles was not accomplished. Dense regions of 10 GPa shock treated and plasma sintered Al_2O_3–10ZrO_2 composite revealed a similar microstructure to the unshocked composite as shown in Fig.6. The average grain size was calculated to be 1.4 μm which is smaller than the grain size of either unshocked or 4 GPa shock treated composites. Microstructural properties calculated from micrographs of Al_2O_3–10%ZrO_2 composites are given in table IV. The highest grain surface areas calculated were those of the shock treated composite at 10 GPa. The fracture mode of

Fig.5. SEM micrograph of 4 GPa shock treated plasma sintered Al_2O_3–10%ZrO_2 composite (5 minutes in H_2 plasma)

5.7 µm
|--------|

Fig.6. SEM micrograph of 10 GPa shock treated plasma sintered Al$_2$O$_3$–10%ZrO$_2$ composite (5 minutes in H$_2$ plasma)

Table IV. Microstructural properties of plasma sintered Al$_2$O$_3$–10%ZrO$_2$ composites *

Shock Pressure (GPa)	Average Grain Diameter (µm)	Average Pore Size (µm)	Grain Boundary Area (mm/mm^2)	Fractional Density (%)
–	1.7	1.4	1015	94.3
4	1.7	0.7	445	92.0
10	1.3	1.4	1440	97.7

* – 5 minutes in H$_2$ plasma

this composite was mainly intergranular although a good interparticle bond is evident from the micrograph.

Shock treatment of SiC$_w$ containing powder mixtures at 10 GPa resulted in fractured whiskers, as shown in Fig.3. A high degree of agglomeration is visible which prevented densification, except in the agglomerates. In contrast to the shocked material, a uniform arrangement of particles and whiskers was observed in unshocked Al$_2$O$_3$–30SiC$_w$ composite. Outer regions of this composite is composed of significantly densified (>90% fractional density) fine grains. Some whiskers that have been pulled out during fracture can also be seen in this region (Fig.2). MECHANICAL PROPERTIES – Table I summarizes the mechanical properties of plasma sintered Al$_2$O$_3$–10ZrO$_2$ composites. Overall, these values are lower than conventionally processed composites. The low values can be attributed to remaining porosity in the structure and incomplete bonding of the particles in some cases. The fine grain size achieved by the plasma sintering process, however, is an obvious advantage to most mechanical properties of interest.

Plasma sintering in O$_2$ yielded the best mechanical properties in unshocked Al$_2$O$_3$–10ZrO$_2$ when compared to other plasma gases. Increasing the sintering time of this composite improved the properties as can be seen from table I. Shock treatment did not have a significant influence on the mechanical properties.

Only hardness measurements were possible in low fractional density Al$_2$O$_3$–30SiC$_w$ and Al$_2$O$_3$–10ZrO$_2$–30 SiC$_w$ composites. These values are given in table II. Longer sintering times decreased the hardness of Al$_2$O$_3$–30SiC$_w$ composites. Shock treatment of the powders resulted in lower hardness values in the plasma sintered Al$_2$O$_3$–30SiC$_w$ composite. A significant hardness increase accompanied higher fractional densities in 4 GPa shock treated Al$_2$O$_3$–10ZrO$_2$–30 SiC$_w$ composite when compared to its unshocked counterpart. This may be due to localized high density regions caused by agglomeration due to the shock treatment.

DISCUSSION

Although the highest fractional density achieved by the HCD-plasma sintering process was about 94% in Al$_2$O$_3$–10ZrO$_2$ composites, this value may be increased to full density by several means. Some of the potential possibilities for this purpose are increasing the plasma power and therefore the sintering temperature, increasing the green density, and passing the specimen through the plasma at a certain speed [11]. Similarly, SiC_w containing composites can be densified better if the whisker content is decreased [25]. Proper selection of preconsolidation processes to minimize agglomeration and to increase the green density is also an important consideration [26] Densification is greatly inhibited by whisker networks and agglomerates around them in high whisker content composites [25]. The effect of agglomeration was pronounced in shock treated powder mixtures in the present study. Even though visible agglomerates were eliminated from the powders in the dry form, agglomeration might still have occured during plastic forming of the slurry mixture due to high surface activity of shock treated powders [27].

The high densification rates in the plasma sintering process are generally attributed to high heating rates activating grain boundary and lattice diffusion before coarsening by surface diffusion can take place [12]. High heating rates are not limited to plasma sintering. Rapid zone sintering [28] and microwave sintering [29,30] are other processes where comparable heating rates occur and high densities are achieved within several minutes as in plasma sintering. A second factor, suggested as the source of rapid densification in Kemer and Johnson's work [12], is the very high sintering temperatures reaching to 1900°C. The present study and similarly Cordone and Martinsen's HCD-sintering study [9] show, however, that this factor is not a requirement for rapid sintering of ceramics, because temperatures employed in the two HCD-sintering studies were only in the range of 1300°C. Nonetheless, very high temperatures may be effective in the final stages of sintering. This argument can be based on maximum fractional densities of 96 and 94% that were attained in Cordone and Martinsen's and the present study, respectively, at sintering temperatures of about 1300°C, compared to full densities achieved at 1900°C [12].

Mechanical properties of plasma sintered composites were not at the expected level, probably due to remnant porosity. However, improved properties can be expected if the plasma process parameters such as the plasma power, translation speed of the specimen through the plasma (which was zero in this study), and sintering gas are selected so that complete sintering is achieved.

SUMMARY

Plasma sintering experiments of unshocked and shock treated Al_2O_3–based composites showed that Al_2O_3–$10ZrO_2$ composites can be sintered up to a fractional density of 94% in several minutes, whereas limited sintering is achievable in 30 vol.% SiC_w containing composites. Increasing the sintering time of whisker containing composites did not increase their fractional densities. Mechanical properties of Al_2O_3–$10ZrO_2$ composites were not as good as conventionally sintered composites which can be attributed to remnant porosity in unshocked and weaker interparticle bonding combined with higher porosity in shock treated composites. Proper adjustment of plasma sintering parameters have the potential of producing composites with mechanical properties surpassing those of composites fabricated by conventional routes.

Shock treatment did not improve the sinterability of the composites and mechanical properties of shock treated Al_2O_3–$10ZrO_2$ composites were not much different from unshocked counterparts except for higher hardness values achieved in shock treated Al_2O_3–$10ZrO_2$–$30SiC_w$ composites. Particle agglomeration due to the shock process was observed to have detrimental effects on the properties. Removal or inhibition of such agglomerates may improve the effect of shock treatment on sintering of Al_2O_3–based ceramic powder mixtures. Further study based on the present results will be conducted in this domain.

REFERENCES

1. B.Edenhofer, pp.181–85 in "Source Book of Nitriding", American Society for Metals, Metals Park, OH (1977)

2. O.T.Inal and C.V.Robino, Thin Solid Films, 95, 195–207 (1982)

3. E.Metin and O.T.Inal, J.Mater.Sci., 22, 2783–88 (1987)

4. N.Y.Pehlivanturk, O.T.Inal, and K.Ozbaysal, Surface and Coatings Tech., 35, 309–20 (1988)

5. J.A.Thornton, pp.19–62 in "Deposition Technologies for Films and Coatings", edited by R.F.Bunshah, Noyes Publications, Park Ridge, NJ (1982)

6. G.J.Vogt, D.S.Phillpis, and T.N.Taylor, pp.203–16 in "Advances in Ceramics, Vol.21: Ceramic Powder Science", edited by G.L.Messing, K.S.Mazdiyasni, J.W.McCauley, and R.A.Haber, The American Ceramic Society, Inc., Westerville, OH (1984)

7. J.P.Pollinger and G.L.Messing, pp.217–28, ibid.

8. C.E.G.Bennett, N.A.McKinnon, and L.S.Williams, Nature, Appl.Sci., 217, 1287–88 (1968)

9. L.G.Cordone and W.E.Martinsen. J.Am.Ceram.Soc., 55, C–380 (1972)

10. D.L.Johnson and R.Rizzo, Am.Ceram.Soc.Bull., 59, 467–72 (1980)

11. D.L.Johnson, W.B.Sanderson, J.M.Knowlton, and E.L.Kemer, pp.656–65 in "Advances in Ceramics, Vol.10: Structure and Properties of MgO and Al$_2$O$_3$ Ceramics", edited by W.D.Kingery, The American Ceramic Society, Inc., Columbus, OH (1984)

12. E.L.Kemer and D.L.Johnson, Am.Ceram.Soc. Bull., 64, 1132–36 (1985)

13. J.S.Kim and D.L.Johnson, Am.Ceram.Soc.Bull., 62, 620–22 (1983)

14. K.Upadhya, Am.Ceram.Soc.Bull., 67, 1691–94 (1988)

15. G.Thomas, J.Freim, and W.Martinsen, Trans. Am.Nucl.Soc., 17, 177 (1973)

16. G.Thomas and J.Freim, Trans.Am.Nucl.Soc., 21, 182–83 (1975)

17. P.C.Kong, Y.C.Lau, E.Pfender, K.McHenry, W.Wallenhorst, and B.Koepke, pp.939–46 in "Ceramic Transactions, Ceramic Powder Science II", edited by G.L.Messing and H.Hausner, The American Ceramic Society, Inc., Westerville, OH (1988)

18. E.K.Beuchamp, pp. 139–74 in "High Pressure Explosive Processing of Ceramics", edited by R.A.Graham and A.B.Sawaoka, Trans Tech Pub., Ltd., Switzerland (1987)

19. O.R.Bergmann and J.Barrington, J.Am.Ceram.Soc., 49, 502–07 (1966)

20. D.W.Richerson, "Modern Ceramic Engineering", Marcel Dekker, Inc., NY (1982)

21. G.R.Anstis, P.Chantikul, B.R.Lawn, and D.B. Marshall, J.Am.Ceram.Soc., 64, 533–38 (1981)

22. A.G.Evans and E.A.Charles, J.Am.Ceram.Soc., 59, 371 (1976)

23. D.B.Marshall, T.Noma, and A.G.Evans, J.Am. Ceram.Soc., 65, C–175 (1982)

24. T.H.Hare, K.L.More, A.D.Batchelor, and H.Palmour III, pp.265–80 in "Materials Science Research, Vol.16: Sintering and Heterogeneous Catalysts", edited by G.C.Kuczynski, A.E.Miller, and G.A.Sargent, Plenum Press, NY (1984)

25. T.N.Tiegs and P.F.Becher, Am.Ceram.Soc.Bull., 66, 339–42 (1987)

26. M.D.Sacks, H.W.Lee, and O.E.Rojas, pp.440–51 in "Ceramic Transactions, Vol.1: Ceramic Powder Science II", edited by G.L.Messing, E.R.Fuller,Jr., and H.Hausner, The American Ceramic Society, Inc., Westerville, OH (1988)

27. J.Golden, F.Williams, B.Morosin, E.L.Venturini, and R.A.Graham, p.72 in "Shock Waves in Condensed Matter", edited by W.J.Wellis, L.Seaman, and R.A.Graham, American Institute of Physics (1982)

28. M.Harmer, E.W.Roberts, and R.J.Brook, Trans. J.Brit.Ceram.Soc., 78, 22–25 (1979)

29. T.T.Meek, R.D.Blake, and J.J.Petrovic, Cer-
am.Eng.Sci.Proc., 8, 861-71 (1987)

30. J.Wilson and S.M.Kunz, J.Am.Ceram.Soc., 71,
C-40-41 (1988)

WEAR RESISTANCE OF PLASMA NITRIDED
HIGH SPEED STEELS

D. Kakaš
University of Novi Sad
Novi Sad, Yugoslavia

M. Zlatanović
University of Beograd
Beograd, Yugoslavia

ABSTRACT

Samples made of high speed steels M2 and M35 were plasma nitrided at various process conditions and tested in "Pin-on-disc" machine. The wear intensity as well as the coefficient of friction have been measured. SEM micrographs taken at the surface of the samples and tools have shown that in the case of complex stresses the intensity of wear can be reduced by applying plasma nitriding process. It has been shown that the service life of tools is dependent on the depth and microhardness distribution of plasma nitrided layers.

THE WEAR OF PLASMA NITRIDED SAMPLES made of HSS was investigated in order to improve the service life of cold working tools for the special applications.

The tools for bacward and forward extrusion which are exposed to compressive stresses up to 3000 MPa during operation, were made of HSS. In this case, the plasma nitriding technology gives a good possibility for improving the wear resistance of tools. In the case considered, the holes made in workpieces had the tolerance of $30 \mu m$. According to the recommendation given in the literature /1,2,3/ the nitrided layers of the depth 19 to 58μm were produced on the surface of the tools and test samples made of steel grade M2. The layers consisted of only diffusion zone improved the durability of tools and the brittleness of the punch head been avoided /4,5/.

In order to avoid a very expensive industrial test, the "Pin-on-disc" vear test has been performed by using the HHS samples.

Considering the wear intensity of the tools for backward and forward extrusion, the most critical part of the tool surface is the tool head edge due to the maximum overflow of the workpiece material. At the same place the very high compressive stresses exist during tool operation. On the "Pin-on-disc" machine a similar conditions can be produced.

EXPERIMENTAL PROCEDURES

The samples (pins) were made of two grades of high speed steels: M2 as the most frequently recommended HHS and M35 cobalt containing steel recommended for very hard workink conditions. The difference in wear resistance of the samples caused by the properties of these steels was found /6/. Both types of materials were quenched and tempered to the hardnes of 62 HRC.

Plasma nitriding has been performed at the Faculty of Electrical Engineering - Belgrade, in the plasma nitriding unit MONO 5 which is equipped by fully automatic control system. Based on the preliminary experience /7/, four groups of plasma nitriding process parameters were applied:

- Type A - temperature 540°C, treatment time 60 minutes and 15% of nitrogen in N_2-H_2 mixture.
- Type B - temperature 480°C, time 15 minutes and 6% of nitrogen.
- Type C - temperature 509°C, time 30 minutes and 9,5% of nitrogen.
- Type D - very intensive sputtering during the ion bombardment heating, time 40 minutes and 10% of nitrogen.

The microhardness profile of the diffusion zone is shown in Fig.1. The depth of nitrided layers was the following: Typ A - 58 μm; Typ B - 19μm and Typ C - 39μm. The Auger electron analysis (equipment PHI model SAM 545 A) confirmed the differences in the nitrogen content from 6 to 9% at.

The disc were made of carburising steel surface treated to the hardnes of 60 HRC. After grinding the disc surface roughness was uniform.

The "Pin-on-disc" test was made at the load F=15N and the speed v=0,2m/s. The wear resistance was investigated at the unlubricated conditions. Wear resistance was measured in function of changes in wear surface diameter. The head of pin has shape of hemisphere with radius 12 milimeters.

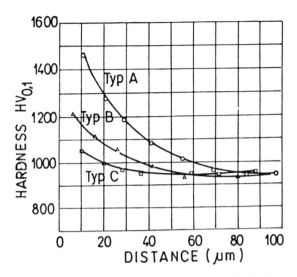

Fig.1. Microhardness profile of nitrided layer after various plasma nitriding cinditions

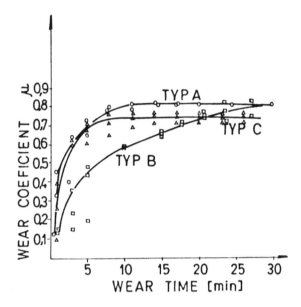

Fig.3. Friction coeficient of the pair:plasma nitrided pin - cemented disc, after various plasma nitriding process on M2 steel.

RESULTS AND DISCUSION

It has been found that for the samples made of the same steel grade the wear intensity is significantly influenced by plasma nitriding process conditions. As an example, the Fig.2. shows the wear resistance as measured by the changes in wear surface diameter at the top of the pin.

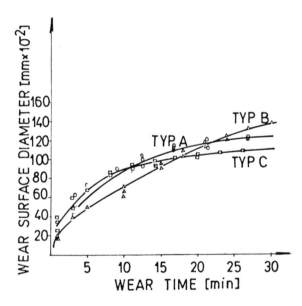

Fig.2. Wear resistance of M2 pins plasma nitrided at various conditions.

The influence of plasma nitriding conditions on the friction coeficient was also found (Fig.3). Across section at the middle of wear surface of pin was shown at the Fig.4. Difference in thicnes of nitrid layer was result of wear process. By control of wear surface diameter on pin,

Fig.4. Across section in the middle of pins wear surface.

142

In the case of the pins plasma nitrided at very intensive sputtering during the ion bombardment heating (process Type D) in plasma nitriding chamber, the scattering of the "Pin-on-disc" results was observed. This is illustrated in Fig.5 and Fig.6.

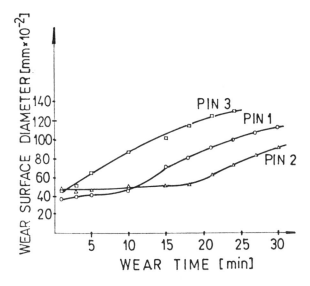

Fig.5. Wear resistance measured at the surface of plasma nitrided pins made of M2 steel (proces Type D).

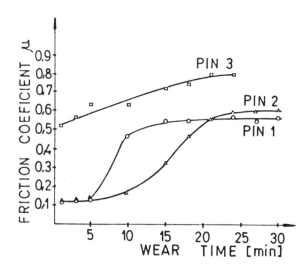

Fig.6. Friction coefficient of the pair: plasma nitrided M2 pins (process Type D) – carburising disc.

The difference in wear behaviour of samples Type A and samples Type B, both made of steel grade M35 was also found (Fig.7). This difference in wear behavior could be explained by the SEM analysis of the pin wear surface. As an example the wear patterns of two samples were compared. Pin Type A has the largest thickness of the nitrided layer and the highest microhardness. The wear surface of pin (Fig.8) is relatively flat (Fig.9) with certain number of holes with various

depth. These holes seem to be result of the carbide ejection during the wear period (Fig.10).

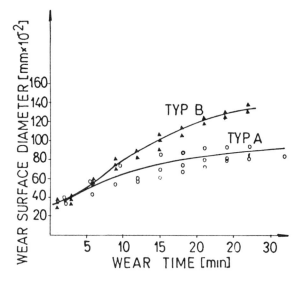

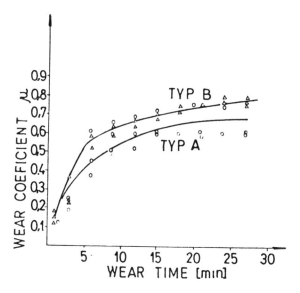

Fig.7. Wear resistance and wear coeficient of M35 plasma nitrided pins after different šlasma nitrided treatments.

In Fig.11 the part of the surface is shown with the carbide particles just to be worn out (arrow).

This result shows that the influence of carbide shapes and dimension, specialy in the thin surface layer, very influence on wear behavior of plasma nitrided pin. Very big influence has also nitrided matrix of pin and conection between the carbide and matrix.

Some of our investigation was concentrated on influence of various parameters of plasma nitriding process at carbide phase in surface layer of M2 and M35 steels /8/.

143

Fig.8. Wear surface (M35 pin - Type A)

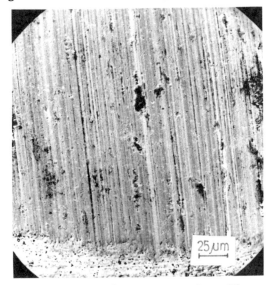

Fig.9. Relatively flat wear surface (Type A)

Fig.10. Holes at wear surface (M35 pin-TypeA).

Fig.11. Place at the wear surface of pin(M35 - Type A) with carbide just to be worn out - signed with arrow.

Pin Type B has the smallest thickness of the nitrided layer and the lowest microhardness. Several surface zones are visible on the pin after the wear test (Fig.12). The first zone (Fig. 13) has a ring form and it is relatively flat (ZONE I), while in the central region (ZONE II) very narrow surface was found. A typical appearance of ZONE II was shown in Fig.14.

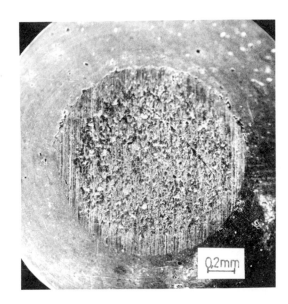

Fig.12. Wear surface on M35 pin (type B).

144

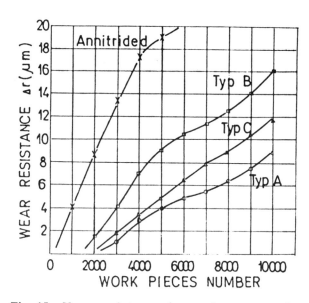

Fig.13. Wear surface of M35 pin (type B) with
two different zone of wear.

Fig.15. Wear resistance for various type of
plasma nitrided punch made of M2
steel compared with annitrided punch.

CONCLUSION

The "Pin-on-disc" test have shown that the
wear phenomena at the contact surface: plasma
nitrided HSS pin and carburised disc, are dependent on a very small variation in thickness
of nitrided layer and its surface hardnes. For
the wear tests performed with no lubrication the
influence of HSS composition was also noticed.

The presence of the carbide particles at the
contact surface strongly influences the wear behavior of the tribological pair considered and for
this case the surface structure with the minimum
size of carbide particles could be recomended. A
fine distribution of carbide particles is also desirable.

In order to increase the wear resistance of
HSS made cold working tools the stress distribution which can suppress the carbide particles to
be worn out is also desirable. Plasma nitriding
can be applied to get the corresponding microhardness and compressive stresses distribution.
In this case, the plasma nitriding process parameters have to be carefully adjusted in order not
to get brittle surface structure.

The optimum plasma nitriding condition have
to be selected for any particular working conditions of the tools for forward and backward extrusion.

Fig.14. Typical appearance of ZONE II at M35
pin (Type B).

These results can be used to explain the
results of industrial test of tools for backward
extrusion made of HSS. As the criterion for the
wear resistance the lowering of the punch diameter (Δr in μm) was used. The results are given in Fig.15 for different types of plasma nitrided layers and compared with not nitrided
punch. This wear resistance were measured for
one typical geometry of punch. In this case for
very anconvenient shape of punch geometry.

REFERENCES

1. Zlatanović M., Osobine površinskog sloja do-
   bijenog postupkom jonskog nitriranja na brzo-
   reznom čeliku Č.6980, JUSTOM 83, Novi Sad,
   1983.
2. Bell T., Dearnley P.A., Plasma surface engi-

neering, First Int. Sem. on Plasma Heat Treatment, Senlis 1987.

3. Lahtin Ju.M.,Kogan Ja.D.,"Azotirovanie stali", Mašinostroenie, Moskva (1976).

4. Kakaš D.,Lupuljev D.,Veselinović Č.,Zlatanović M.,Influence of the ion nitriding parameters on cold forming tools quality, 5 Int. Con. on Heat Treatment, Budapest 1986.

5. Kakaš D.,Lupuljev D.,Zlatanović M.,Contribution to investigation of plasma nitriding on tools for cold working, First Int. Sem. on Plasma Heat Treatment, Senlis 1987.

6. Hoyle G., "High Speed Steels", Butherwoth (1988).

7. Kakaš D.,Zlatanović M.,Rac A., Istraživanje uticaja plazma tehnologije na rezultate habanja brzoreznog čelika, JUSTOM 89, Vrnjačka Banja 1989.

8. Jordović B.,Kakaš D.,Zlatanović M.,Prilog istraživanju izmene strukture površinskog sloja plazma nitriranog brzoreznog čelika, JUSTOM 89, Vrnjačka Banja 1989.

PLASMA (ION) NITRIDING AND PLASMA (ION) NITROCARBURIZING UNITS, APPLICATIONS AND EXPERIENCES

Wolfgang Rembges, Jens Lühr
Klöckner Ionon GmbH
Leverkusen, FRG

Plasma (Ion) Nitriding
and
Plasma (Ion) Nitrocarburizing

Units, Applications and Experiences

by W. Rembges, J. Lühr,

5090 Leverkusen 3, FRG

I. Introduction

Plasma (Ion) nitriding (PN) and plasma (Ion) nitrocarburizing (PNC), better known as Ionitriding[1], has been used in the past in a wide range of applications [1] which is constantly increasing [7]. Due to economic and environmental considerations Ionitriding is rapidly replacing the atmospheric processes like gasnitriding or gasnitrocarburizing which use high amounts of gas or the salt bath processes with their associated poisonous problems for the environment [20].

The enhanced properties of ionitrided workpieces are already proven in numerous industrial applications. Ionitriding has the flexibility [12] to solve many tribological problems :

      wear
      fatigue
      corrosion
      adhesion or friction.

[1] Ionitriding, registered trademark of Klöckner Ionon GmbH, Leverkusen, FRG.

The properties of the nitrided layers are influenced by the material, prior heat treatment, carbon content, the content of uncombined nitride precipitations forming elements (Chromium, Aluminum, Molybdenum, etc.) [14] and process parameters like:

      time
      temperature
      gas composition
      treatment pressure

To achieve these advantages in a contract heat treatment shop or in a factory under series production conditions certain requirements have to be up to standard for the applied installations like:

      flexibility
      reliability
      efficiency
      environmental

Different designs of plasma nitriding units are available, but their suitability to the above mentioned requirements have to be discussed critically. In the literature for example it is stated that furnaces with thermal insulation combined with pulsed power supplies give a better temperature distribution and lower energy consumption. To disprove that statement this paper will describe the actual standard of PN/PNC-units and will give some information about the engineering properties of the different furnace systems. Thus it gives a better understanding of the process, about the most favourable system and will help anybody in his own decision.

The effect on series or daily production with different load sizes will be described as well as some engineering points, such as how to make an efficient and reliable plasma nitriding unit with sufficient good temperature distribution and acceptable energy consumption.

Several thermocouples (normally up to four) are controlled by a microprocessor control board [3], to regulate the heating rate to the temperature as well as to control the desired temperature uniformity within a load. All setpoints and actuals are

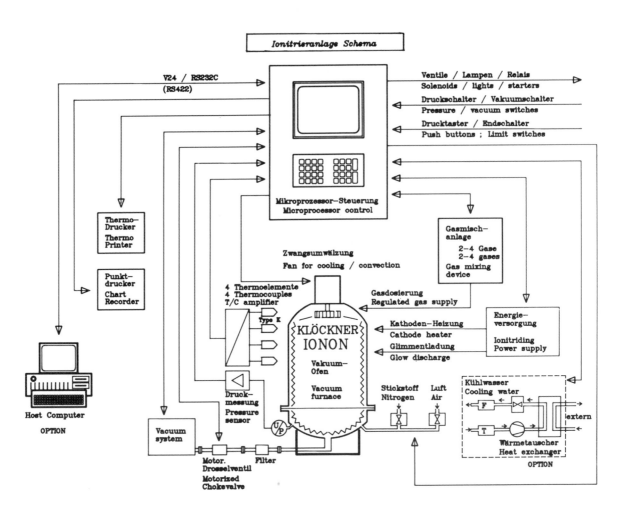

Picture 1: Lay-Out of a multifunctional Ionitriding installation

II. Todays Plasma (Ion) Nitriding Units

Modern Ionitriding-units consist of various components as given in an overview in picture 1. The furnaces are normally equipped with a suitable sized vacuum pump which guarantees a short evacuation time and a sufficient low gas flow rate of treatment gas during the process. The heating up process can be influenced by integrating an indirect heating system which reduces the cleaning time by the glow discharge (sputtering).

recorded, printed and can be documented by a host computer or PC via a RS232 serial port. An operator screen leads through the program steps.

The electrical power to heat up the workpieces and to create the glow discharge around the workpieces is produced by a power supply. The size of the power supply has to be chosen according to the load size and furnace dimensions. SCR (thyristor bridges) controlled units range up to 1000 KW. Their reliability is already proven in more than 150 units.

II.1. The Effect of Installation Designs on the Ionitriding Process

A Ionitriding cycle consist of the following steps:

> unloading and loading
> pumping down
> heating
> nitriding treatment
> cooling

The unloading and loading operation has to be performed according to the workpiece requirements. The arrangement of a load influences the temperature distrubition because in a vacuum of less than 10 mbar there is no convection; only radiation takes place. In addition, care must be taken in the arrangement of the workpiece supports, their contact areas being minimized to reduce shadowing and masking effects.

After pumping down initial heating of the load can be performed in plasma heat treatment furnace by using three different systems or a combination of each:

> a) convection
> b) radiation
> c) direct Plasma heating

a) Convection:

The heating by convection is performed in a vacuum furnace slightly under atmospheric pressure in pure nitrogen atmosphere. The gas is heated by resistance heating elements and is circulated by a fan through the furnace. Important for the heat transfer from the heating elements to the workpieces is the heat-transmission-coefficient of the gas, the heat capacity of the gas and the speed of the gas molecules along the heaters and the workpiece surface. In order to prevent a too big heat loss to the furnace wall, the chamber must be carefully insulated, otherwise the gas will loose too much heat to the furnace wall and will not be able to heat the load efficiently. (The influence of the insulation on the glow discharge conditions is discussed in chapter II.4.). The efficiency is reduced with decreasing difference in temperature between the gas and the workpieces. The uniformity within a load is very much influenced by the similarity of the gas stream throughout the load.

b) Radiation

The heating by radiation is performed in a vacuum furnace at low pressure (≤ 100 torr) where there is no significant convection. The load of workpieces to be heated has to be surrounded by heated surfaces. These heated surfaces may be heating elements, which are electrically heated or could be an internal shield (cathode heater) which is heated by a plasma of the glow discharge (principle see c)). Three important points in the heat transfer from the heating elements (or internal shield) to the workpieces are the temperature differences between the load and the heaters and the ratio of the surface areas of the load to the heater and their emission coefficients.

With decreasing temperature difference the efficiency of the heat transfer is reduced. Normally the principle lay-out of the cathode heater (it surrounds the whole load) gives a higher surface area, while the effective area of the resistance heaters is comparatively low ($\leq 50\%$). Therefore the cathode heater has a higher heat transfer to a load compared to resistance heaters.

c) Direct Plasma Heating

Direct plasma heating is performed in a vacuum furnace at low pressure (≤ 10 torr) in a glow discharge. The heat is produced at the surface of a workpiece (or cathode shield). The workpiece has to be connected to the negative pole (cathode) of a power supply and a voltage drop of several hundred volts causes electrons to move towards the anode. The gas molecules are ionized as a result of collisions with electrons. The positively charged ions of high kinetic energy move towards the cathode and their energy is converted into heat at the surface.

Due to the difference in specific mass between an electron and an ion the velocity of the electrons is much higher and there is a build up of ions near the cathode. Thus the ionic intensity is largely independent of the distance between the workpiece and the anode. By controlling the total current of a plasma power supply the energy which is going into the surface of the workpiece is regulated. That means the heating up speed can be directly controlled. Thus the heat transfer from the plasma into the surface is a function of the power density and the thermal conductivity of the material of the workpiece.

Compared to the other methods mentioned the highest heat transfer can be achieved with direct plasma heating, because it is independent from the actual temperature. The limitation of the heat transfer is given by physical conditions like an electrical arc. In this event an automatic control system with a very sensitive circuit breaker is required to prevent an arc.

The most efficient system is a combination of system b) and c). In system b), the cathode heater, is used in the lower temperature range and the direct heating is used at temperatures of more than 300°C (570°F) up to the treatment regulation point. Thus already during the heating period a depassivation and activation of the surface is performed.

Having reached the treatment temperature different systems of furnaces and power supplies are applied to maintain the glow discharge and to perform the nitriding process. The effect of these systems are connected to several physical conditions which limit the applicability.

II.2. Comparison of Different Systems

This will be done by comparing results, which are published in the literature [6,9] and are achieved under series production conditions in the automotive industry. The following example, shown in picture 2, represents a load, weighing 6000 kg (13200 lbs) and consisting of 3800 gears. The effective ionitrided area is 70m² (750sqft). This Ionitriding furnace has been in production for more than 5 years, 5700 hours per year. The down time has been less than 4% which demonstrates the reliability of the used system.

The quality of automotive components is controlled by statistical methods, which allow to estimate the capability of a system [4,6]. A reliability study of the surface hardnesses of statistically chosen and tested ionitrided gears has been performed. The results, produced in the above mentioned equipment and named with unit

Picture 2: Ionitriding bottom loader, integrated in a series production line
$\emptyset_i$: 1800mm (6ft), H: 1800mm (6ft)

"A", are summarized in picture 3 which shows the distribution of the hardness, taken at the pitch of seven different sized gears, in a histogram (upper part) based on data getting from probability paper (lower part). The standard deviation ±3σ of the hardness of appr. 3800 gears, total weight 6000 kg (13200 lbs) is only ±42HV. These figures are not corrected by the influence of the core hardness of the material which has been supplied from five different forging companies in a 28 to 32 HRc heat treated condition before machining. This result describes the capability and reliability of the used system.

In comparison to these results, hardness values from another plasma-nitriding treatment, performed in a thermal insulated furnace with a pulsed power supply and convection heating are shown in the same picture 3 and designated unit "B".

While the effective nitrided area is only 25 m² (270sqft) with unit "B" compared to 70 m² (750sqft) with unit "A", it is to the first attempt not understandable, that in spite of the smaller surface area the deviation in hardness within ±3σ is ±230HV, which is five times higher and unacceptably high. The average hardness (50%-value) is ≈90HV less, compared to the results achieved in unit "A" , but the maximum measured hardness is in the same range in both cases.

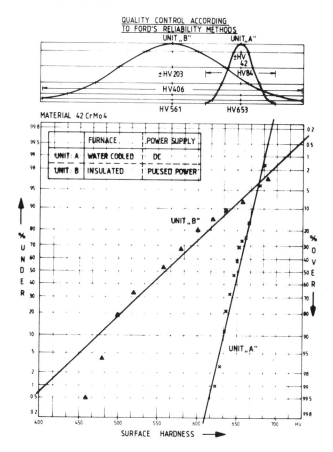

Picture 3: Hardness reliability study of two different workpieces and systems after PN-treatments

The results received in unit "A" correspond with the values which are normally possible. This is already demonstrated in various reports from different authors [2,5,8]. Therefore an explanation has to be found, why in unit "B" the difference between the lowest and the highest value is much bigger, than in unit "A".

A characterization, given by the authors of the article of unit "B", can only explain a few of the possible

reasons [9]. The same influences described, coming from the base material 42CrMo4V (AISI 4140), are valid for the workpieces, treated in unit "A" [4,5,6]. Therefore other parameters have to be taken into account, to explain the big differences. According to the experience of the authors of this article the design of the equipments may be responsible. This experience corresponds to J.P.Lebrun [19] who mentioned that 80 to 85% of his production is successfully performed in furnaces with watercooling. Comparing both units, two items - 1) the different power supply systems and 2) the distinct heat losses due to furnace designs - must be taken into consideration.

II.3. The Effect of Power Supplies

The development of the plasma(Ion) nitriding process in the past thirty years was very much influenced by the various possibilities of building power supplies of different designs. Investigations about the influence of the type of power supplies on the plasmanitrided layers, which were investigated some 15 years ago, have been repeated recently using pulsed power supplies [21].

The description of the effect of power supplies on the nitrided layer can easily be done by comparing the surface hardnesses as a function of the power density, received with different systems. Picture 4 shows surface hardness as a function of power density, received after nitriding of the same material with different power supply systems. It is obvious that a minimum power density is necessary to achieve a constant result in surface hardness. The results have been achieved under reproducible conditions in a specially designed installation. Special attention was paid to the way of measuring the power density, because the normal analog instruments (for voltage and current) may lead to a misreading of the real power density.

This result corresponds to former but unpublished results of the authors showing that the minimum density is much less comparable to data published in paper [10].

Hardness = f (Powerdensity)
Material: 42 CrMo4
Gas: N2/H2 ... 25/75 %

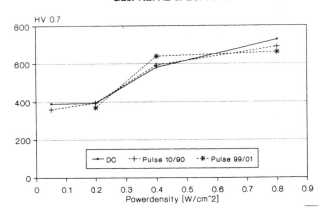

Picture 4: Hardness results after PN-treatments with different types of power supplies as a function of the power density

The various types of power supplies do not show any significant influence on the hardness of the plasmanitrided material throughout the whole range. No influence on the used different ratios of the duration of the pulse or the pause between the pulses has been detected.

This result shows that the differences in the plasmanitriding results, mentioned in picture 3 must have a different reason.

The characteristic of the relationship between hardness and power density gives an indication of the hardness being influenced whenever the density is not high enough. The next chapter will describe how the power density is influenced by the design of the furnace.

II.4. Influences of the Furnace Design

The heat loss in a vacuum furnace is only influenced by the radiation of the workpiece temperature to the furnace wall. Because of the low applied pressure of less than 10mbar gas convection gives no influence to the heat loss. Thus the energy consumption

is mainly a question of the design of the furnace wall, respectively of the wall temperature.

Investigations about the influence of different wall temperatures on the power density are shown in picture 5. It is illustrated by five furnaces with different volumes of up to 6 m³ (200cft). They are loaded with workpieces to a constant load density and it can easily be seen the bigger the furnace volume, the lower is the power density on the workpiece area. Different furnace wall temperatures are created by different furnace designs; insulated or cooled to a certain amount makes an influence, too.

Powerdensity = f(Furnacevolume)
at different furnace wall temperatures

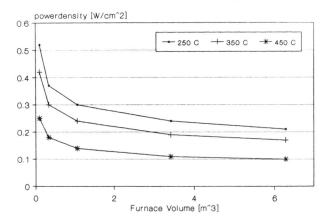

Picture 5: Change of power density at different wall temperatures and furnace volume, constant load density in each furnace

Smaller furnaces will automatically have a higher power density than bigger ones. Therefore insulation of bigger furnaces will result in an obviously too low power density. Smaller furnaces may be insulated.

Summarizing both effects it becomes obvious that with increasing furnace wall temperature and useful volume the effective power density is reduced. Knowing that under a certain value of power density the optimum hardness cannot be achieved, the high spread in hardness of unit "B" becomes understandable.

An increased quality without changing the design of unit "B" can only be achieved by reducing the number of workpieces. Thus the production would be reduced which is not acceptable. A more economical way would be to increase the energy consumption by a different furnace design (less insolation or more efficiently cooled furnace wall, for example by water).

III. Demonstration of Felexibility and Reliability

The following pictures show several typical loads which have been treated in an Ionitriding furnace with 1800 mm (6 ft) useful height and a loading diameter of 1800 mm (6 ft) (see picture 6). It has been chosen because it is the most economical one [11].

Picture 6: Standard Ionitriding Installation Type HZI 1800/1800 G360 M

The furnace volume is appr. 6m³ and has a water cooled furnace wall, is equipped with an indirect glow discharge heating system and has a quick cooling system. To reduce the heat loss two heat shields are connected to the inside of the furnace wall. The water cooled wall acts mainly as a heat exchanger at the end of the

cycle when the furnace is backfilled with nitrogen and the gas is forced through the load by a fan. The whole nitriding cycle is microprocessor controlled; the power supply on the right hand side has a maximum power of 360 KW, while the average consumption on temperature is about 100 KW. The maximum power may be used during heating up only.

The total engineering and design of an Ionitriding equipment and the wide range of functions available in the Ionit[2] microprocessor software gives a high production security and makes it very easy to set operating conditions by the operator. The software accepts up to 24 programs to be stored for different applications [3]. Integration in series production lines has been done.

An application of such an equipment, shown in picture 7, are automotive seat-rails, used by one of the famous German automotive companies.

Picture 7: Load appr. 3000 seat-rails during loading and unloading of an Ionitriding furnace

An Ionitriding cycle is performed automatically for a load of more than 3000 seat-rails. They are nitrocarburized to a compound layer thickness of ≥5μm and a surface hardness ≥350 HV0.5. For more than five years this

²)Ionit, registered trademark of Klöckner Ionon GmbH, Leverkusen, FRG.

treatment has been performed without any interruption and with constant quality properties.

Whenever possible, for instance during the weekend, this equipment is used for other components and longer cycle times. Picture 8 shows a smaller load of pump-bodies as they are used with booster pumps. The material is grey cast iron. In this application the energy consumption is much less, compared to the above mentioned load, because the furnace load is less than half of the designed capacity. Thus the equipment works still profitable [11].

Picture 9: Load of cylinders, total weight appr. 6000 kg (13200 lbs)

Picture 8: Load of booster pump bodies after PN-treatment in a HZI 1800/1800 Ionitriding (PN) furnace

The same furnace (see picture 9) is also used for other types of workpieces like highly stressed barrels, which are used in road construction machines. This load has a weight of about 6000 kg (13200 lbs). The material is an Aluminum free nitriding steel. The energy consumption is approximately the same as with the parts, shown in picture 7, but the surface area is less.

Mostly the loads consist of workpieces of different sizes. Picture 10 demonstrates by a load of appr. 500 workpieces of different geometries the flexibility of the Ionitriding equipment and the process itself. Steels with low carbon content as well as steels which are higher alloyed like stainless or maraging steels are plasmanitrided in the same furnace. Only the process parameters like temperature, time or gas composition have to be changed. Even different materials are treated at the same time.

The above presented examples give a brief impression what is possible with Ionitriding. In all cases the requested quality is reproducible received. The quality of nitrided layers may be different from those mentioned above, whenever other systems than the Ionitriding system are applied.

Picture 10: Load of about 500 different workpieces after an Ionitriding treatment, useful hight 1800 mm (6 ft)

IV. Conclusions

Ionitriding processes like plasma (Ion)nitriding (PN) and plasma (Ion) nitrocarburizing (PNC) are already used in a wide range of applications and are known as progressive nitriding treatments. Their tribological behaviour and advantages are already known. Using these processes for daily production, the product quality can be influenced by the engineering and design of the furnace. Thermal insulation of the furnace reduces the flexibility of the process and is limiting the surface area to be plasmanitrided or plasmanitrocarburized, hence the profitability. Modifications of power supplies do not help in this case.

The gasnitriding as well as the salt bath nitriding technologies are working at atmospheric pressures, or slightly overpressure. Thus certain expensive security provisions have to be included in the production areas.

Plasma (Ion) nitriding (PN) or plasma (Ion) nitrocarburizing (PNC) are vacuum technologies which do not require these expenditures. They are applied more and more in heat treatment, because they have an industrial standard and the potential to substitute atmospheric processes.

V. Literature

[1] Hombeck, F.: Das Ionitrierverfahren, Berg-und Hüttenmännische Monatshefte 116 (1971) 11, p.484 - 491

[2] Rembges, W.: Einfluß der Wärmebehandlung auf das Nitrierverhalten von Vergütungsstählen, ZWF 73 (1978) 6, p.329-332

[3] Rembges, W.; Oppel, W.: Mikroprozessorsteuerung für die Wärmebehandlung Gas Wär. Int. (1984) Bd. 33 Heft 6/7, p.349-353

[4] Rembges, W.: Randschichthärten von Getriebezahnrädern IN- LINE-Prozeßß, VDI-Z Bd 127 (1985) 19, p.771-774

[5] Weck, M.; Schlötermann, K.: Plasmanitriding to enhance gear properties, Metallurgia No.8, August 84, p.328-332

[6] Roelandt, A.;, Elwart, J.; Rembges, W.: Plasma nitriding of gear wheels in mass production, Surface Engineering, 1985, No. 3

[7] Hombeck, F.: Forward View of Ion Nitriding Applications, Proceedings of an Intern. Conf. on Ion Nitriding Cleveland, Sept.1986, page 169

[8] Rembges, W.: Fundamentals, Applications and Economical Considerations of Plasma Nitriding, Proceedings of an Intern. Conf. on Ion Nitriding, Cleveland, Sept.1986, page 189

[9] Zimmermann, M.: Application de la nitruration ionique au reenforcement des vilebrequins, Trait. Thermique, 1987, No. 215, p.54-59

[10] Knüppel, H.; Brotzmann; K., Eberhard, F.: Nitrieren von Stahl in der Glimmentladung, Stahl u. Eisen 78, 1958, p.1871-1880

[11] Hombeck, F.: Scientific and economic aspects of plasmanitriding, Proceedings Volume, Shanghai 1983

[12] Rembges, W.: Moderne Plasmatechnologien und -Anlagen für die Wärmebehandlung von Bauteilen, Thermo Prozess-und Abfalltechnik,Vulkan Verlag Essen,Page 181-186

[13] Rembges, W. : Plasmanitriding of PM Parts, MPR November 1988, p.765-768

[14] Hombeck, F. Rembges, W.: Plasmanitriding of Transmission Components ,Practical examples and results of series and single productions, Proceedings of Int. Seminar on Plasma H.T., Senlis ,France (1987), p.315-328

[15] Seyrkammer, J., Klingemann, R., Rembges, W.: Möglichkeiten des Einsatzes von plasmanitrocarburierten Sinterbauteilen, HTM 43 (1988) 6, page 348 - 353

[16] Schlieper, G., Manolache, V., Rembges, W.: Mechanical and wear properties of ionitrided Cr-Mo sintered steel, presented: PM-Exhibition, July 7-11, 1986, Düsseldorf

[17] Elwart, J. : Berücksichtigung des Plasmanitrierens bei der Konstruktion, ZWF. 73 (1978) 12, S. 641 - 647

[18] D.B. Pat. P 29 15 983.9-52

[19] Lebrun, J.P.: Technical developments and industrial applications of ion nitriding, Proceedings of Int. Seminar on Plasma H.T., Senlis ,France (1987), p. 425-444

[20] Bell, T., Dearnley, P.A. : Plasma Surface Engeneering, Proceedings of Int. Seminar on Plasma H.T., Senlis, France (1987), p. 13 - 53

[21] Hombeck, F., Oppel, W., Rembges, W. : Plasma(Ion)Nitriding and Plasma(Ion)Nitrocarburizing, Its Units and its Applications, Proceedings of the 1. Int. Conf. on PSEE, DGM (1988), p. 277 - 287

Authors

Dr. rer. nat. W. Rembges, Head of Process Engineering and Process Developement Dept. of Klöckner Ionon GmbH, 5090 Leverkusen 3 / FRG Director of Engineering, Klöckner Ionon of America, Charlotte, N.C.

J. Lühr, Process Engineering and Process Developement Dept. of Klöckner Ionon GmbH, 5090 Leverkusen 3 / FRG

INDUSTRIAL ADVANCES FOR PLASMA NITRIDING

Reinar Grün
Plasma Technik Grün GmbH
Postfach 210643
D-5900 Siegen, FRG

Abstract

In this paper some industrial experiences for the use of a plasma nitriding process will be presented. For the understanding of the process handling with some simple but impressive models the difficulties by the treating of single parts or mixed workloads are explained. For somebody interested in plasma nitriding plants some guide-lines for the use and the selection of a suitable process and device are given. Especially by the use of a pulsed dc power supply many of the industrial problems could be solved.

Introduction

THE PLASMA NITRIDING, also called glow discharge nitriding or ion nitriding is a well known technology for many years. Special features compared to the other technologies like salt bath nitriding or gas nitriding led to special applications for this technology. In the past the industrial use had a difficult standing on the market, because this technology was expensive and difficult to handle.
Some of these problems were as follows:

- handling of process by skilled people only

- big workloads with reproducible results

- temperature distribution within the workload

- overheating of small parts

- surface demage by arcing

- hollow cathode handling

To understand the reasons it is useful to know something about the process not only on a scientific base but also on an industrial practise base. For this it is useful to have information about the background of the special plasma behaviour. For the user of plasma plants it is necessary and helpful to understand the background of the plasma handling in the industrial practise.

Plasma Nitriding Technique

FOR THE PLASMA NITRIDING the following equipment is needed:

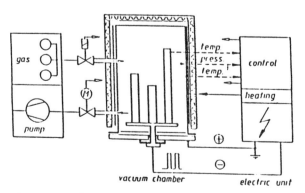

Fig. 1: Plant for PULSE PLASMA diffusion treatment

- vacuum vessel with pumping system to get the gas pressure of about 10 to 1000 Pa (approx. 1 - 10 Torr)

- gas supply for nitrogen containing gases

- electric power supply for a glow discharge

- electric power supply for seperate heating

- process control unit

The parts to be treated are electrically insulated from the surrounding chamber and connected to the dc power supply. The workload will form the cathode (-) within the anothic chamber (+) The electric positive ions in the gas will be accelerated to the negative workload. By this bombardment the parts are heated up and the ions are able to form nitrides at the surface.

The specific voltage to current characteristic has to be considered for the handling of the plasma process. (1)

To get good and reproducible results for the workpieces within a complete workload and for different workloads it is necessary to have a plant with which it is possible to secure the same conditions for every treatment. This sounds very simple but in many cases this cannot be realized and secured.

For the handling of a plasma nitriding plant the following parameters of such a device influence the results of a treatment:

- Temperature of workload

- Time for treatment

- Gas mixture in the vessel

- Gasflow through the vessel

- Pressure in the vacuum vessel

- Applied plasma voltage

- Plasma current density

For the plasma nitriding process all these parameters are not independent from each other and have different correlations and interactions together. On the other side it will be the best if it would be possible to control and regulate all these parameters seperately and independently. It is necessary that the special plasma parameters like vacuum ressure, electric voltage and the current densitiy can be fixed and the temperature of the workpieces can be regulated seperately.

At the first conference on Ion Nitriding in Cleveland 1986 the advantages of the pulsed dc technology for a power supply were presented and explained (1). With this technique many problems of the conventional dc technique in the past could be solved.

To try to clarify several reasons for the above listed problems, some correlations between the different parameters will be explained by the following examples.

If the temperature of the workload and the vessel is low or if the vacuum pressure in the vessel is high or both together, the plasma density is high and the reaction of the process will be as follows:

* Voltage/current density range for the abnormal glow discharge, which is necessary for a proper process, is small. This leads to:

  - stability of plasma difficult
  - high risk for arc formation

* The current density onto the workpiece will be high, therefore:

  - heating of the workpiece fast
  - sputtering rate by ions high

* The glow discharge is narrow and leads to:

  - good covering of surface contours
  - good penetration into holes and slots

Normally the third piont is desired but the problems of the first mentioned point are too difficult to handle in many cases.

The opposite situation of these points with low plasma densitiy in the case of high temperature or of low vacuum pressure is as follows:

158

* Voltage/current density range for abnormal glow discharge wide:

  - stability of plasma good
  - low risk for arc formation

* Low current density:

  - heating of workpiece slow
  - sputtering rate by ions low

* Glow discharge extended:

  - covering of surface contours bad
  - penetration into holes and slots bad

For the easy handling of the process it is desired to have this situation especially during the heating up period. Due to the low temperature at the beginning of a process it has to be started at low pressures with the mentioned advantages and disadvantages. In the case of heating up the workload by plasma only the heating time will be very long, compared to the heating by an auxiliary heating.

By changing the pressure in the vessel it is possible to control the surface covering by the glow discharge. In fig. 2 it is shown that the covering of a surface at complicated geometries is much better at the high pressure

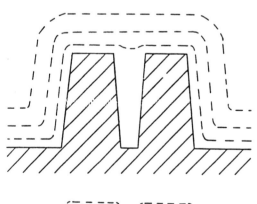

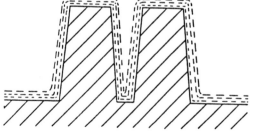

Fig. 2: Glow discharge at a workpiece
top: low gas pressure
bottom: high gas pressure

mode. On the other hand, if desired, it is also possible to prevent a penetration into holes and slots by using a lower pressure. Between these two situations a hollow cathode effect has to be considered which is difficult to handle and to control.

Models for heating up of workpieces by plasma

BY THE PLASMA PROCESS each workpiece is covered by a glow discharge and will be heated up by the ion bombardment at the surface. The heating will be higher if the surface area is big. Each workpiece has a behaviour like a single heating element in the workload. In the following, by different models with simple geometries the practical situation will be explained.

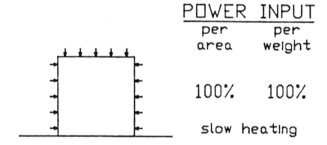

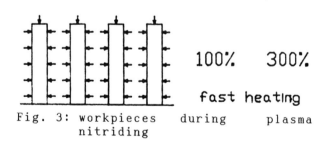

Fig. 3: workpieces during plasma nitriding

In fig. 3 two examples are compared with each other.

At first it is shown a compact workpiece ("fistlike") with a low ratio of surface to volume or mass. In this case the heating up of the mass by plasma through the surface, as shown by small arrows, is supposed to be 100 %.

In the second part of fig. 3 different pieces (like fingers or flat hands) are shown with the same total mass as in the first part. Corresponding to this, the total surface of these pieces, faced to the plasma is much higher. Therefore in this model the energy input will be about 300 % in relation to the compact piece so that the heating will be faster

In the most cases it is rather simple to run the plasma nitriding process with a workload of equal or similar parts. The heating up by the plasma will be the same for every part. Therefore it is aspired to put together only similar parts into the vessel. But normally in the daily business, especially in a heat treatment shop this is impossible because a furnace has to be loaded completely every day with mixed workloads to earn his money.

A more realistic but also more difficult situation will be the above two seperated fist and sheet or finger like cases together within one part (fig. 4). In such cases the heating by plasma only will be a problem at any time because the risk for overheating of the upper thin part is very high. By this the tempering with the lost of case hardness and distortions could happen very easily.

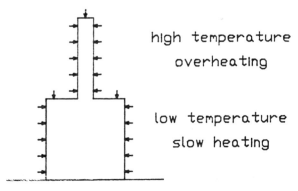

high temperature
overheating

low temperature
slow heating

Fig. 4: Workpiece with different geometries at the same piece

The hollow cathode effect is a special situation of the glow discharge in a hole of a workpiece. The main influences on this effect are the size of the hole, the gas pressure and the current density. Into such a hole the discharge can penetrate and the gas in it will be overheated up to extremely high temperatures. If the overall energy input per area could be reduced, the influence of the hollow cathode effect will be much lower.

In all the above cases it will be helpful to run a furnace with seperate and from the plasma heating independent heating unit. This seperation is possible by using a pulsed dc power supply for the plasma with a repetition frequency of more than 1 kHz and with a variable ratio of the on- and off-periods.

A further step to reduce the above mentioned problems will be the use of a hot wall furnace compared to a cold wall system. Due to the balance of heat input and output the temperature profile in the workload will be much better as shown in fig. 5.

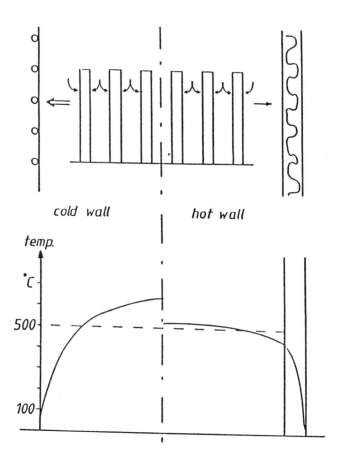

Fig. 5: Cold and hot wall chambers with workload and temperature profiles

Examples

FOR THE ABOVE SHOWN MODELS of workpieces and workloads some examples are given below. The experiences with these examples were made in a treatment center by using plasma nitriding plants with hot wall vessels up to 2 m high and an outside auxiliary heating with an insulation as shown in fig. 1 and fig. 6.

The power supply for the plasma has a pulsed dc characteristic with a repetition frequency up to 16 kHz and a pulsed power up to 200 kW.

In fig. 7 an example for a dense workload of similar parts is shown. One single part of such a gearshift lever (fig. 7a) has a length of about 200 mm. A complete workload of about 1000 pieces are positioned on 6 levels with a package density as shown in fig. 7b. This high package density was possible only by using a hot wall furnace to avoid overheating in the centre of each level.

An example of a mixed workload is shown in fig. 8. Here we have very thin fingerlike pieces (0,3 kg/unit) together with rings for the support of cutter blades (40 kg/unit).

An example for a complex geometry as mentioned in fig. 4 is shown in fig 9. Here we have thin parts on the outside of this cutter blade so that the overheating of these parts (fig. 9b) have to be avoided.

Conclusion

THE PLASMA NITRIDING PROCESS as a modern technology for the surface treatment of metals had several difficulties in the past, due to the specific handling know how necessary. If somebody is interested in a plasma nitriding plant he should know something about the process and the background of the parameter correlation. By this he can caluclate the best solution for his special needs. Some points to prepare such a decision are as follows:

- handling of the device with similar or mixed workloads

- heating and cooling times of a workload to optimize the complete process time

- electric energy consumption of a complete unit for specific workloads

- effective and economical using of a plant as a single or tandem unit

- effective and economical using of the vessel volume with a suitable workload density
- control and recording unit for the completeprocess, clear and expedient

Big dimensions of a vessel or high values for power supplies are not good and not necessary in any case. It will be not economical e.g. if in a big vessel about 20 % of the volume would be used only due to problems of the temperature distribution.

A big power supply e.g. will be not economical if it has to feed the cooling device of the vessel.

Finally it will be recommendable to harmonize different components of a complete plant like power supply, vessel, pumping system and gas supply. Otherwise the advantages of one component could not be utilized completely.

Reference

(1): R. Grün, Pulse Plasma Treatment, the innovation for Ion Nitriding, Proc. Int. Conf. Ion Nitr., Cleveland, 1986, Editor T. Spalvins, ASM 1987

Fig. 6:
Plant for pulsed plasma nitriding
with hot wall vessel
pulsed power: 200 kW
vessel: diam: 700 mm, hight: 2000 mm

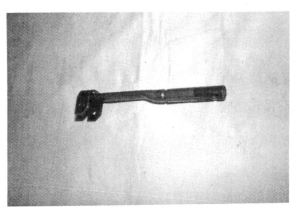

a

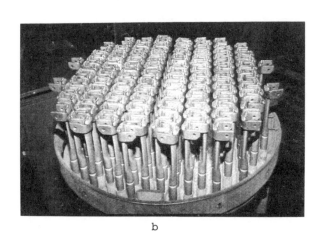

b

c

Fig. 7:
Workload of approx. 1000 gearshift
levers in a dense package

Fig. 8:
Workload of cutting tools (40 kg each)
together with thin pins   (0,3 kg each)

a

Fig. 9:
Cutting tool with fingerlike geometry
outside

b

ION NITRIDING ALUMINUM EXTRUSION DIES

Paul H. Nowill
Nitron, Inc.
Lawrence, Massachusetts, USA

Aluminum extrusion dies are an ideal candidate for ion nitriding. In Europe and Japan most extrusion dies are ion nitrided. Yet, gas nitriding dominates the U.S. market. Failure in the U.S. market has been caused by past marketing mistakes, a perceived lack of importance and inertia. Successful commercialization of ion nitriding will require overcoming these three obstacles.

Market History

Ion nitriding was first introduced into the U.S. market as a commercial service in the late 70's. The equipment, like its German counterpart used ion plasma as its source of heat. Heating extrusion dies in this way requires skilled operators in order to prevent overheating parts of a die while underheating other parts. Inexperienced operators failed to properly heat some dies and ion nitriding received a reputation for spotty coverage. Modern equipment has overcome this problem, but the reputation remains.

Technical Summary

In aluminium extruding a ram forces heated aluminium through dies to form long strips of complex shapes. The pressure and high temperature requires tough dies that hold their shape and maintain tolerances. The abrasive action of the aluminium requires hard surfaces that do not spall or wear out.

Extrusion dies are usually made of H13 steel and heat treated to a hardness of 48 to 52 HRC. H13 is selected for its toughness and its ability to withstand heat. Ion nitriding adds a dense, smooth and ductile surface of gamma prime nitrides (Fe_4N) with a hardness of over 70 HRC supported by a hard zone containing diffused nitrogen. In service gas nitrided dies last 5 times as long as intreated dies. Ion nitrided dies last 7 to 8 times as long. In addition, the "hard layer ductility" of ion nitrided dies allows them to be run 20% faster without spalling.

The Ion Nitriding Process

Extrusion dies are ion nitrided by placing them into a vacuum furnace, heating to about 950° F., adding nitrogen and hydrogen gases, and applying an electric potential. The gases become ionized and ions bombard the surface of the dies. Molecules of metal sputter off, react with the ion plasma to form nitrides, and redeposit on the die surface. All but a thin layer of these nitrides decompose to release nitrogen into the diffusion zone.

Temperature and ion conditions must be maintained at proper levels during nitriding. Extrusion dies have holes of various shapes and sizes. Aluminum flows trhrough these holes and it is their walls that need strengthening and hardening.

The ions heat all exposed surfaces. However, the walls of holes get hotter than other parts of the die for two reasons; (1) opposing walls radiate heat back and forth and (2) plasma electrons generate secondary electrons and extra heat by bouncing against the walls before they escape to the chamber wall. This extra heat is easily carried away to the rest of the die except when the die has thin fingers of metal. (Thin fingers may not conduct enough of the excess heat to the main part of the die to prevent overheating of the tips of the fingers.) Unfortunately, many extrusion dies have these fingers.

Overheating fingers is a problem in ion nitriding extrusion dies, and fear of doing so may cause operators to lower process temperature and risk spotty coverage. But this overheating does not occur if the power density is kept low. Power desnity can be kept low by having a large surface area in the furnace or by adding radiant heat to maintain process temperature.

Some Examples

Exhibit 1 shows two typical extrusion dies. If these two dies are processed together in a small load the die on the left is likely to be underheated and the die on the right is likely to be overheated. Even the most skilled operator would have difficulty processing this load without radiant heaters.

If the load is increased to fill the chamber and carefully fixtured, the power density per die can be lowered and both dies can be safely processed without radiant heaters, but a full load may consist of 40 or more dies. Radiant heaters give the flexability of running small loads which gives the ability to offer fast turnaround.

Example 1

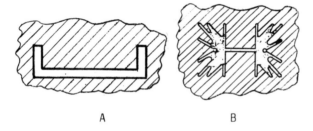

A   B

When "A" and "B" are processed in the same run, "A" is likely to underheat and "B" is likely to overheat.

The Surface

Ion nitriding creates a dense monophase surface of Gamma Prime (Fe_4N). In addition, ion nitriding does not precipitate nitrides in the grain boundaries below the nitride layer. The dense monophase surface and the nitride free difussion zone of ion nitriding increases fatigue strength and resists spalling better than gas nitriding surfaces. Ion nitrided surfaces do not spall.

Marketing of Ion Nitriding

Ion nitriding is the accepted technology for nitriding in Europe and Japan. In the U.S. many aluminum extruders believe that gas nitriding is good enough and that ion nitriding is no good because it gives spotty case. If die life is the only consideration, then gas nitriding is "good enough" since many dies become obsolete before they wear out. The extra die life from ion nitriding is not worth the added cost and risk of spotty case.

Why did ion nitriding succeed in Europe and Japan and not in the U.S. market? One reason is that ion nitriding is more energy efficient, but more importantly, ion nitrided dies can be run 20% faster than gas nitrided dies. Ion nitriding was sold to the U. S. market for increased die life, which is unimportant if the dies have not been wearing out.

Speed of service is important in the U.S. market, so much so that some large users have their own gas nitriding equipment even though it is not cost effective (purchased service, gas or ion nitriding, is much cheaper).

Value Analysis of Ion Nitriding

The aluminum extruder will make his purchase decision on some form of value analysis. One of the reasons for the marketing failures of the past was that the marketing focus was on die life, which is very important in most markets, but it seems not very important for aluminum extruders. Table 1 shows the assumptions of a sample value analyses based first solely on die life, then on capital cost, and finally on total system cost.

Table 1

Assumptions

Die cost	$ 300	
Gas nitriding cost	37.50	1.50 $/lb.
Ion nitriding cost	75.00	3.00 $/lb.
Number of dies	300/year	
Equipment cost	$ 2 Million	
Capital charge rate	15 %	

Ion nitriding,

Improves life   50 % (over gas nitriding)

Increases speed 20 % (higher lubricity and less galling)

Savings on dies

Assumption 1: All dies run full life

Die cost savings (300 x 50%) = $ 150.00

Less added ion nitriding cost   37.50

Savings per die   112.50

Yearly savings (112.50 x 300) = $ 33,750

If you are selling ion nitriding, this is the assumption you would like to use. But your customer might respond with the next assumption.

Assumption 2:   No die wears out

Added yearly cost of ion nitriding:
(300 x 37.50) = $ 11,250

If the customer does not see die wear as a problem, ion nitriding looks like unnecessary added cost. Conclusion: Gas nitriding is good enough. Even with a realistic assumption of die life, the savings based on ion nitriding look trivial.

Assumption 3:   replace 15% of gas nitrided and 0% of ion nitrided dies

Added yearly cost of new dies
(.15 x 300 x 337.50) = $ 15,189
Less added ion nitriding cost   11,250

Yearly savings $ 3,939

Conclusion: This saving is not worth the risk of trying something new. If the assumptions are realistic, it will be very difficult to sell ion nitriding to alluminum extruders on the basis of die life.

Savings on Capital and Labor

Capital savings:

(capital x capital charge x 20 percent productivity increase)

($2,000,000 x .15 x .20) = $ 60,000/year

The full capital "savings" is obtained only when the equipment runs at capacity, but the need for new capacity will be delayed. It may be imposible to buy capacity a 20% increment. Switching to ion nitriding may delay the need for a two million dollar new investment.

Labor savings:

The labor savings will be less than 20 percent, but there will be savings both in operating the extruder and in maintaining the dies. For this example a savings of $ 30,000 is used.

Quality and turnaround improvement:

The quality of the surface finish of the extrusions can be maintained much longer without tuch-up of the dies. The dollar value of this quality improvement is hard to measure.
The 20% shorter run times reduces turnaround time. If the plant is running near capacity the reduction in queuing time can be substantial.

Savings on capital and labor:

$ 60,000 + 30,000 = $ 90,000

Less the cost of ion nitriding   11,250

Net savings   78,750

The capital and labor savings is 20 times the savings in die costs (15% assumption). Even if these savings are much less than this estimate, they will still be much larger than the savings in die cost.

Marketing strategy

The first step in good marketing is to find out what the customers want. A 50% to 100% increase in die life does not make the sale. Ion nitriding's true value is its dense, ductile and high lubricity surface which allows extruders to run their machines faster. This translates into cost free increases in capacity and productivity, which are far more valuable than increases in die life.

Customers also want fast service. It is not cost effective for extruders to do their own nitriding, so why do they do it? Either they are not aware of the full cost of their inhouse gas nitriding, or the faster inhouse service is worth the extra cost. We have found that customers will pay double for fast turnaround for about 30% of their work. (Our costs are more than double because we have to run partial loads to give this service.) This willingness to pay higher prices suggests that fast turnaround, not ignorance of the true cost is the reason for inhouse gas nitriding.

Ion nitriding can turnaround faster than gas nitriding because it is a faster process. However this advantage is often wasted by waiting to fill a large furnace. Equipment vendors like to sell large furnaces because their profits are larger, but small furnaces provide the flexibality needed for fast service. Also, ion nitriding shops need to be organized toward faster service.

There is no need for inhouse gas nitriding by extruders. The quality of ion nitriding is so superior that extruders with gas nitriders should scrap them and either buy ion nitriding services or get an ion nitrider.

As commercial ion nitriding shops we need to demonstrate to extruders that they can get good service and fast turnaround from us.

ION NITRIDING CHALLENGES WEAR

John N. Georges
Plasma Metal S. A.
Luxembourg/Luxembourg G. D.
L-1817

Abstract

Wear or in particular wearparts are an expensive aspect in today's industry. It may be accepted that nearly in all branches of industry the wear problem is very up to date.
Primary there are the costs of such replacement parts. The most expensive part of a manufacturing facility is the shut down time of a production line. Do not forget the man hours for replacement and adjustment work.
Using ion nitriding, brings indirect profits of a non negligible dimension. It has a great importance in the challenge of total quality parts. Indeed absence of wear and absence of excessive tolerances of mechanical parts in a machine or tools, allow to reach unequalled top quality.
Many parts have to function under quite severe conditions.
Plasma nitriding is the heat treatment that might be adapted best to the industrial requirements. All the treating parameters are flexible, and most of them can be instantly changed during processing.
It has been proven that a close coordination with the "end-user" brought best results.
In this regard, the design of the parts, the correct choice of the material, previous heat treatment, etc. helped to bring up top quality parts.
The author shows with some industrial applications, and step by step, explains the way he went, to reach the final goal.

WEAR PROBLEMS are still contributing substantially to general costs in the manufactoring industry. Any worn part results in maintenance breaks, new set-up time and accordingly to lost production. Furthermore, any required and replacement part, spare part or tooling component has its own cost. Today's industrie is very competitive and any further cost saving potential is important and has to be considered for maintaining a strong position in the marketplace. Improvements are often related to initial investment (better materials and additional process costs), but show considerable cost savings at the bottom line. Therefore it is sometimes required to take one step back to provide a platform for three steps forward. Ion nitriding is a process which provides substantial wear resistancy, improvements to a wide range of materials and is considered as most reasonable (by process cost) as compared to other known new process technologies.

Ion nitriding is meanwhile a well known surface hardening process. It is not the purpose of this paper to enumerate the advantages and disadvantages of plasma or ion nitriding.

We want to show this time how a product or a so called wear part can be improved, together with the assistance of the customer's maintenance people.

At first we shall see some examples of a wire plant. Steel cord is a severe antagonist to mechanical parts during its processing. Starting up with the dies, the draw cones, etc....

As an example let's take a set of drawpulleys for steel cord.

steel cord

At the beginning the customer tried out nearly all the possibilities on the material's side. Starting with cast iron, hardened cast iron, flame hardened steel, case hardened steel, but the lifetimes of these pulleys were moreless the same but poor.

The lifetime was dependable of the product as well. Single wires did not damage the system as much as steelcord.

The decribed parts had to be replaced every three to four months.

After a close up view of the problem, we made a set of pulleys of D2 steel. After grinding and polishing, we did a vacuum quench.

The vaccum quench brought us a hardness of 56 - 58 HRC.

We did some repolishing on the pulleys and started a pulsed ion nitriding cycle of 16 hours.

This is how we did it:
- the pulsed ion nitriding was considered as the third tempering of the parts after quenching.
- as D2 steel contains 12% of chromium we heated up the parts with hydrogene and high voltage pulses to depassivate the parts.
- after one hour maintaining time at 535 degrees C we introduced nitriding gas mixture for another 16 hours at 520 degrees C.

The result showed an increase from 57 - 59 HRC of the core hardness and a value of +/- 1200 HV-0.3 on the nitrided surface.

These first two pulleys are now working for about three years, totalling aproximatively 300,000 km of steel cord without a trace of wear.

There is another point to be outlined, which is the financial aspect. It is true that the price of such a set of pulleys is at least three times hig-

her than for instance similar parts which are just case hardened and ground. Compared to the performance, the price can be neglected.

In the same wire plant we could improve the lifetime of the cones of a draw bench, while the costs of manufactoring came down. The cones were initially out of D3 steel, quenched, tempered and polished. It was a very hard work to polish the cones after quenching and tempering as for proper operation of the cones, a mirror like polish is required.

We successfully replaced the D3 by prehardened nitralloy. We machined the parts, ground them and before ion nitriding we made a super polish on them.

We nitrided the cones for about 42 hours at 520 degrees C. At this time it was absolutely important to use pulse plasma equipment to prevent damage on the super polished surfaces by arcing.

After plasma nitriding, a short repolishing brought the initial shine back.

The lifetime of the cones is about three times the lifetime of the D3 parts whereby the costs of the nitralloy parts are 25% lower.

drawbench with drawcones

Finally before leaving the wire-plant there is one of the most interesting parts in regard of the wear resistancy. This part is very small, as you can see it on the sketch.

This part is a wire guide in the wire twister. There are two guides built on a disk which rotates around central wires to make a cord.

These wire guides do not move and the wire itself slides over them without any lubrication.

It is useless to say that wear is the big problem in this area of the wire twister.

Nearly everything has been tryed out:
Plasmasprayed alloys,
Plain tungstencarbide guides,
TiN coated guides,
Plasma nitrided guides, made out of nitralloy.

By making different tests it has been proven, that the plasma nitrided nitralloy guides gave best results. Although normally one would have expected another result.

It was very interesting to discover, why things went that way.

The plasmasprayed and the plain tungstencarbide wire guides had the same wear effect. Both surfaces were very hard but porous.

The steelcord while passing over that surface broke out a particle and carried it along over the whole contacting surface making scratches and while doing so, the wire picked up more and more particles that way.

Similar to how a diamond is shaped by using diamond powder, these wire guides destroyed themselves in the same way.

Which regards the TiN coated wire guide that was worn out quickly we believe the TiN coating was extremly thin and the backing material probably too soft.

The plasma nitrided wire guide had the better lifetime by far and which is very interesting for the user, once the part is worn out it can be reground and re-ion nitrided at low cost. In average such a part can be reused that way up to eight times.

There are many other maintenance managers which have got their own problems in other fields of industry. We shall have a closer look at a ceramic facility. Unfortunately these people did not want to show all the parts where we have been successful in lifetime improvement as this would have disclosed some of their production secrets, but finally we could get permission to publish results of one machine. We are in the Villeroy and Boch factory in Luxembourg, which is famous for their ceramic plates and cups,etc....

On the picture we see a roller as they call it. With this roller, the underside of a plate is quickly shaped and polished. The roller is rotating at high speed and cuts into the backside of a plate to shape it. There is a special angle between the roller and the plate.

The customer used to make the rollers out of normal carbon steel with some hard chrome plating. The problem with the hard chrome plating was to get an equal plating all over the roller and into it's profile as well.

The advantage of this process is according to the customer's opinion,that one can remove the hard chrome plating for doing some rework on the roller itself and to make a replating before reuse of the roller. The same roller made of nitralloy and ion nitrided was unfit for rework because of its hardness. Even better lifetime couldn't persuade the customer to use our technology as he insisted in having the possibility for reworking availibility on his rollers.

But one should never give up. The problem was solved in using 4140 steel instead of nitralloy. Pulseplasmanitriding provided the same layer all over the roller. The lifetime is equal to the hardchrome plated system. A rework is possible and all this at a lower cost.

Another interesting problem could be solved on a labelling machine at a mineral water company in Belgium. These rollers are built on a rotary shaft which picks up paper labels and forwards them on glue over to a brush system which labels the bottles.

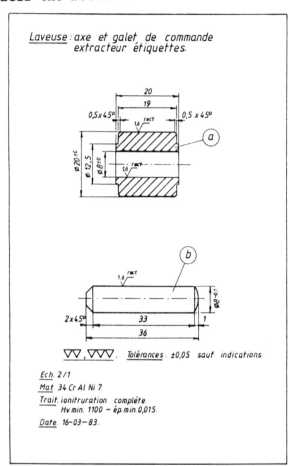

Laveuse : axe et galet de commande extracteur étiquettes.

$\nabla\nabla$, $\nabla\nabla\nabla$. Tolérances : ±0,05 sauf indications.

Ech. 2/1
Mat. 34 Cr Al Ni 7.
Trait. ionitruration complète.
    Hv.min. 1100 – ép.min. 0,015.
Date. 16-03-83.

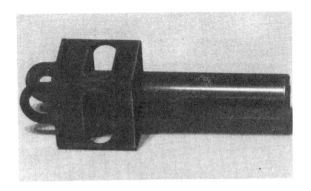

Another part which has its application in space research and was built to test accelerations up to 400 g. Ion-nitriding was choosen to prevent seizing. After testing, the part was bent but it did not seize.

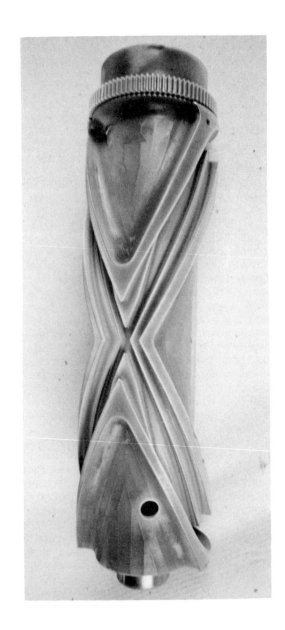

This business is based on glue and water. When the roller gets excessive clearance on its pin the paper labels are no longer picked up properly. Useless to state how the bottles look like after labeling.
Many possibilities have been tested.
Teflon, nylon, bronze, brass bushings and needle bearings have been tried out, but without great results. Knowing the superior seize resistance and low friction between two ion nitrided surfaces and additionnaly by the resistance to corrosion, we made rollers and pins out of nitralloy. After a nitriding of the outside diameters,
we made an extra nitriding of the bore by a hollow cathode.
Since assembly of the ion nitrided parts the unit works since 12 months without any problem.

This is a part which is up to three feet high made exclusively of nitralloy. One can see all the sliding slots and a gear on top of it. This part when it is built on a machine moves fast and works like a high speed yankee screw. As perfect lubrication cannot be guaranteed all the time, therefore the choice has been made in the same direction like previously the roller and pin of the labelling machine. The part is used successfully and we got a reorder from the customer.

Another field of lifetime improvement are the so called pellet plates. Very often we found pellet plates made of H13 steel or similar. After first wear signs these plates can be reground and ion nitrided. Here as well it is comfortable to use pulsed DC power because of the many holes which can "fire." There will be no markings around the holes which should be rather sharp after the regrinding operation. The pressure of the vessel should then be kept below or very close to one millibar. Normal nitriding time is at our place about eight hours after heat up to 520 - 530 degrees Celsius.

We tried this out the first time with the Eurofloor Company which is the leading manufactorer of plastic floor covering material. The original plate was worn out and we got it in for regrinding. After regrinding the lifetime was about one week running time. Everybody was disappointed. We got it back for another regrinding and I made a pulse plasma ion nitriding treatment on that pellet plate. I never got the exact composition of the steel, but the hardness I finally measured was about 1250 HV 0.3. This time the pellet plate lasted for 20 months and they then had to destroy it, as they got an electrical heater problem and the whole pellet machine was sticking togethter with platstic material in it.

Plasma nitriding is a fine process for heat treating tools, like milling tools, boring tools, deep hole boring tools, etc... mostly one could claim every kind of tool which uses tungstencarbide inserts. The tools can that way be easily manufactored out of aisi 4140 or similar and after total completion they are plasma nitrided. That way one gets no distortions and the wear resistance against the chips is good and the friction is low.

To conclude I would like to outline that in nearly all cases,
it is very important to get a complete follow up of the parts in the facility and it must be possible as well to trace the parts whenever you want to do so.

Finally a perfect help is given to the engineer, if somebody can tell how the parts to be lifetime improved worn out. Which material has previously been used,which heat treatment has been applied, and especially what have been the results as well on the product's quality as on the life cycle time of the part.
I still want to insist that it is so important that the quenching operation of the steels is carried out properly and accordingly to the steel manufactorer's specifications. If you want to be successful in liftime improvement of machine parts thanks to the help of plasma nitriding, it is essential to have best possible quality material before nitriding. I am convinced that nearly everyone in this room got parts which apparently were well hardened, even the" Rockwell" hardness tester showed up with good figures, but the microstructure of the material was that bad, after an incorrect quenching operation, that the part was finally unfit for use.

I hope that this short paper gave you some good guidelines and contributes to some new ideas regarding applications of the process. If some of your customers, or even you,experience wear problems on equipments, and would like to know if ion nitriding would be suitable, just feel free to contact me. Thanks to fax transmissions, any drawing, details or specifications can be communicated. Any addditional costs for correctly applied ion nitriding are a wise investment, since cost savings through the extended life cycle of parts are many.

ION NITRIDING STELLITE

Paul. H. Nowill
Nitron, Inc.
Lawrence, Massachusetts, USA

Stellite is a high performance alloy with outstanding resistance to wear and corrosion. Stellite can be both surface hardened and age hardened in an ion nitriding chamber.

Stellite is used as an alternative to nitrided tool steels, but performance can be further improved by nitriding stellite. Since stellite is expensive, the performance requirements that justify its use probably justify ion nitriding as well.

Description of the Stellite samples

Stellite is a cobalt based alloy. The samples, known as alloy 6B, consisted of 30% chromium, 4 to 5% tungsten, 1% carbon and a combined total of 10% nickel, iron, and other elements leaving roughly 55% cobalt. When the stellite rod was cut into samples, the surfaces work hardened to 62 Rockwell C (at 100 grams). The core hardness was 38 RC. In un-nitrided stellite chromium and chromium carbides form a hard phase distributed throughout the softer matrix of cobalt. Tungsten is added to increase the strength of the material.

Stellite is solution heat treated at 2250° F. It is not quenched but it can be age hardened by holding it at 1500° F. for three hours.

Ion nitriding of Stellite

Since stellite has only 3% iron, iron nitrides do not play a significant role in hardening. But chromium is a strong former of nitrides. Chromium nitrides are probably the source of hardning in the nitriding process. The role of nitrogen dissolved in the cobalt matrix was not investigated.

The tests used a nitriding process identical to that used on steel for a gamma prime surface of "average" thickness, i.e. a temperature of 950° F., a gas of 25% nitrogen and 75% hydrogen, and a time of 8 hours.

These tests produced surface hardness of about 72 RC at 300 grams. A number of valves and bushings were processed in these tests and they are performing well in service. The tests on corrosion resistance are not complete, but so far no reduction in corrosion resistance has been observed.

Age hardening Stellite

Since stellite age hardens, it might be possible to increase core hardness and surface at the same time in a nitriding chamber. To test this idea a sample was processed for three hours at 1500° F. in a nitriding plasma. The core hardened to 42 RC as expected but no increase in surface hardness was observed.

The next test included an age hardened stellite sample with steel parts in a normal nitriding run. All the steel parts, including the stainless steel parts, hardened as expected, but there was no increase in hardness on the stellite sample.

The surface of the sample was ground off and it was included in another run. This time the surface hardened. During the age hardening step some kind of surface contamination seems to prevent nitriding.

Next, an argon plasma was used during the age hardening step to prevent, or remove, any contamination. The parts were then cooled to the nitriding temperature, 950° F., and nitrided. This time the stellite sample age hardened and surface hardened.

Conclusions

Stellite is a good material for nitriding. It can be included in a mixed load with other steels in a typical nitriding process. The surface hardness achieved is roughly 72 RC. There is no observed reduction in corrosion resistance.

Stellite can be core hardened and case hardened in an ion nitriding chamber in a two step process. The first step is to core harden at 1500° F. in an argon plasma. And the second step is to surface harden at 950° F. in a nitriding plasma.

ION NITRIDING OF COLD FORGING DIE ALLOYS

P. C. Lidster, G. Pigott
Exactotherm Limited
3115 Kennedy Road
Mississauga, Ontario, Canada, MIV441

Abstract:

The response to Ion Nitriding of four popularly used low alloy forging steels was studied. After preliminary studies leading to the establishment of operating parameters of gas composition, gas flow, pulse power frequency and duty cycle, the surface hardening and diffusion response of these alloys to variations in time, and temperature were studied.

After evaluation of the laboratory data, hammer forge dies produced from one of the alloys were Ion Nitrided and the behavior in production evaluated.

Introduction:

Since its introduction to the Ion Nitriding process in the early 1980's, Exactatherm has endevoured to apply the process to many facets of the Canadian metalworking industry. The acceptance of the Ion Process now as part of the production and process technique, and no longer a novelty or high technology toy, bears witness to the dedication of engineers and technologists from many industries conducting hundreds of trials and tests, and applying the results either to solve technical problems or lower the cost of production of metal or plastic components. Exactatherm is proud to have assisted in over five hundred such trials co-operating with companies ranging from space agencies to local machine jobbing shops, with alloys ranging from exotic titanium alloys to mild steel. The following are some examples of industries which have benefited from the intruduction of Ion Nitriding:

	INDUSTRY	COMPONENT	MATERIAL
1)	Injection Mould Machine	Distributors	4140
2)	Plastic Mould	Compression Mould	P20
3)	Plastic Mould	Large Injection Mould	P20
4)	Stamping Industry	Large Draw Dies	D2
		Extrusion Punches	M4
		Stamping Punch	M2
5)	Forging	Hot Work Dies	H13
6)	Aluminum Extrusion	Extruder Dies	H13

In 1987 Exactatherm was approached by a company in the Hammer/ Drop Forge industry. The company specialized in the shallow impression forging market, and was open to any suggestions to increase productivity, lengthening die life, and subsequently reducing down time.

The alloys popularly used in this industry all contain nitride forming elements in different degrees, Chromium, Molybdenum, Manganese, and therefore should respond very well to Ion Nitriding.

Because of the severe mechanical stresses associated with hammer forging, gas nitrided dies either failed prematurely through cracking, or gave very inconsistent results. Cracking and spalling of the multi - phase white layer associated with gas nitriding was probably the cause of the premature failure.

The Ion Nitriding response of the four most popular forging steels used in Canada was studied. Field testing of one alloy was observed over a period of one year.

Proceedure:

I. Materials:

Samples of four alloys were obtained from the steel producers.
The alloys were in the Hardened and Tempered condition as follows.

ALLOY	CORE HARDNESS (RC)	COMPOSITION				
		C	Cr	Ni	Mo	Mn
1	42/46	0.45	1.50	0.4	0.2	0.90
2	44/45	0.55	1.05	0.85	0.4	0.90
3	37/40	0.50	0.90	0.90	0.35	0.50
4	40/45	0.56	1.10	1.7	0.50	—

Samples were ground to give a uniform surface finish.

II. Ion Nitriding:

A   The furnace used was an Elactec Ionitrider, equipped with auxiliary
heaters, Honeywell DCP Microprocessor, mass flowmeter gas control,
Bell - Jar style. The furnace was powered by a Walker Power " Fox "
unit rated 50KW, pulse power supply.

B   Standard Operating Conditions:

Leak - up rate   —   Max. 10μ/hour.
Starting pressure   —   20 micron min.
Operating pressure   —   3 Torr
Gas flow   —   5 C.F.H.
Gas composition   —   75% H_2 /25% N_2

C   Ion Nitride procedure:

Samples were placed on mild steel hearth, accompanied by two test
thermocouples.

After evacuation to (20) microns, the A.C. power took the samples up to the
operating temperature. The D.C. glow is introduced at this stage in an ionized
hydrogen glow at 500 microns. After a short period to allow for any sputtering, the
reaction gas was introduced and the pressure taken up to 3 Torr. After the prescribed
time, the parts were cooled down to room temperature under recirculating nitrogen
atmosphere. The pieces were then sectioned, mounted, examined microscopically and
micro-hardness testing performed.

III. Ionitriding Of Hammer Dies Produced From Alloy (2):

The forging dies weighed approximately 700 Lbs each. The first step in preparation was to clean the dies in a vapor degreaser. Areas not needing nitriding were stoped off, using the appropriate stop-off paint. The dies were then ionitrided, following very closely, the procedure established for running the samples. After Ion Nitriding, the samples the samples run with large dies were examined for microhardness profile and white layer morphology. The dies were then released for field testing.

The performance of dies that had been built up with weld material and subsequently Ionitrided were also examined.

Results:

Examination of the response to Ionitriding by the four steels show that they follow classic behavior of lower alloy steels.

1. Case depth is proportional to the square root of time ($D \propto \sqrt{t}$) for any one temperature.

2. Case depth is heavily dependant on temperature. An optimum temperature of 950° was choosen for field trials because this temperature gave the optimum combination of surface hardness and case depth.

3. The alloys with the highest alloy content exhibited the highest surface hardness (alloys 1 & 4). Alloy (3), with the lowest chromium content exhabited the lowest hardness, but the deepest penetration.

All alloys responded well to Ionitriding, achieving hardness of between 15 to 20 points Rockwell C above the core, with depths of between 0.013" to 0.020" were achieved in 24 hours at 950°F.
(See Figure I)

Field results were extremely successful. Die life increased from a minimum of 80% to a maximum of 300%. These results have been repeated a minimum of six times. Similar increases were experienced with the heavily welded dies.

Conclusion:

The shallow impression forging industry will benefit considerably from a full program of Ion Nitriding of the popular die steels. A study will now be under taken to investigate the application of Ion Nitriding to medium and deep impression forging dies.

180

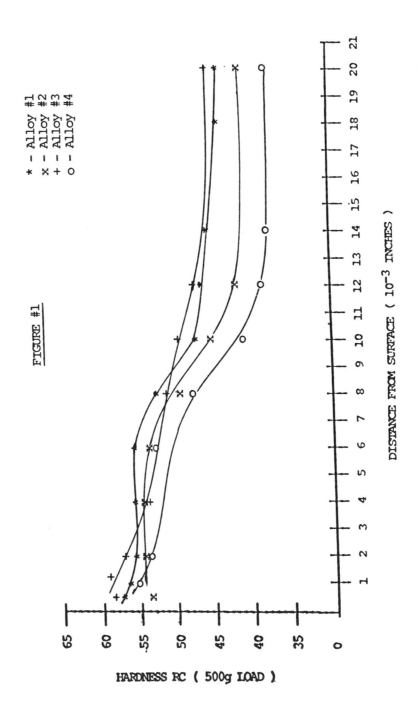

FIGURE #1

* – Alloy #1
x – Alloy #2
+ – Alloy #3
o – Alloy #4

DISTANCE FROM SURFACE (10^{-3} INCHES)

HARDNESS RC (500g LOAD)

PLASMA PARAMETER CONTROL FOR INDUSTRIAL SITUATIONS: THE ROLE OF HOT WALL PLASMA NITRIDING FURNACES

Sidney Dressler
ELTRO
R. D. #8, Box 288
Meadville, Pennsylvania, 16335, USA

PLASMA NITRIDING

Plasma nitriding is being applied to a large variety of manufactured products to improve their corrosion and wear resistance, to increase their fatigue strength and to increase their superficial hardness. When process considerations will permit a selection from competing coating technologies, three technical criteria are generally employed for selecting the most suitable deposition system for a particular application. These criteria include the requirement for a strong bond between the deposited materials and the product surface that can support high localized stress, the need to deposit the coating at a low operating temperature to retain maximum core properties, and the need for deposition on complex product shapes, including shapes with functional side and bottom surfaces and with narrow slots and deep holes.

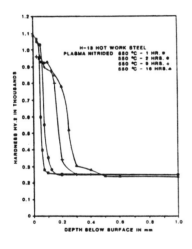

Fig. 1 H13 Hot Work Steel Hardness Profile

Plasma nitriding produces a strongly bonded, nitrogen compound zone and a diffusion zone, actually integral with the original product surface. The nitrogen is added, beginning at the surface and extending inward with a measurable, decreasing concentration gradient as shown in Figure 1. This assures a desirable, gradual transition of mechanical properties from the product surface to the core. This is in contrast to technologies producing overlay coatings simply held by adhesion, where materials are added on top of the product surface, permitting an abrupt change in mechanical properties at the interface.

These diffusion-zone and overlay-coating technologies are not always exclusive. For example, on many manufactured products that must be extremely hard and will be subjected to highly concentrated loading stress, current practice includes first the production of a hard plasma nitrided diffusion zone to more effectively support the second, and even harder, TiN overlay coating.

A plasma nitrided zone is produced by thermal diffusion at processing temperatures safely below the tempering temperature of the product material. Alternative technologies, can operate at much lower product temperatures by using directed high energy beams that have a fundamental line-of-sight requirement. Ion implantation, for example, can produce similar nitrogen concentration gradients. However, in most applications, plasma nitriding using thermal diffusion is less capital intensive, and with no line-of-sight limitation it permits more uniform treatment of all product surfaces and more efficient treatment of large bulk loads containing many small separate parts.

GLOW DISCHARGE FUNDAMENTALS

The glow discharge used for plasma nitriding occurs when a external voltage is applied between two electrodes, positioned within a gas mixture at some suitable partial pressure, as shown in Figure 2.

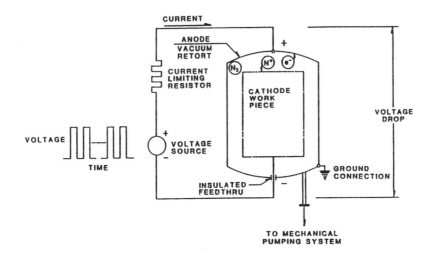

Fig. 2 The Electrical Circuit For a Glow Discharge

One electrode, called the anode, is the vacuum retort, electrically at ground potential. The other electrode, called the cathode, is the work piece to be plasma nitrided. The work piece is connected to operate at a negative potential with respect to the grounded vacuum retort. The voltage source in Figure 2 supplies a variable voltage pulse. With this equipment arrangement, the current in the external circuit can be measured as a function of the voltage drop between the anode and the cathode.

In operation, the space between the vacuum retort and the work piece is filled to some partial pressure with a gas mixture selected for the process. The glow discharge occurs when molecular elements in this gas mixture are ionized by collisions with electrons traveling from the work piece-cathode to the vacuum retort-anode under the influence of the applied electrical voltage. Ionization of the partial pressure gas mixture permits a sustained electrical current, i.e., a negative electron flow from the work piece to the vacuum retort and, more importantly, a positive ion flow from the ionized gas mixture to the work surface being treated.

The newly formed ions will be accelerated toward the product where they

184

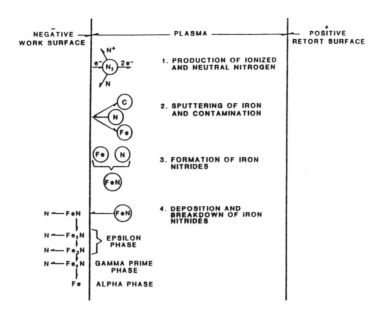

Fig. 3 Surface Reactions During Plasma Nitriding

can combine with the chemical elements at the surface as shown in Figure 3. If the partial pressure gas mixture is predominately nitrogen, the product surface can be nitrided in accordance with the following reactions.

1. Production of ionized and neutral nitrogen atoms by energetic electrons.

    $$e^- \rightarrow N_2 = N^+ + N + 2e^-$$

2. Sputtering of Fe and contaminants from the work surface by these ionized nitrogen atoms.

    $N^+ \rightarrow$ Work Surface = Sputtered Fe and Sputtered Contamination

3. Formation of iron nitrides by the sputtered iron atoms and neutral nitrogen atoms.

    Sputtered Fe + N = FeN

4. Deposition and breakdown of FeN on the work surface.

    $$FeN \rightarrow Fe_2N + N$$

185

$$Fe_2N \rightarrow Fe_3N + N$$

$$Fe_3N \rightarrow Fe_4N + N$$

If the gas mixture is a hydrocarbon, the product surface can be carburized.[1]

THE NORMAL GLOW

This voltage/current density relationship shown in Figure 4 is fully described in the literature.[2] Important physical effects occur when the voltage/current density characteristic enters the normal glow region. In this operating region, a visible glow will cover a portion of the product surface. This visible glow is evidence that process gasses are being ionized and can support the nitriding operation. As the current density is increased, the area covered by the glow will increase until the entire product surface is finally covered with the visible glow.

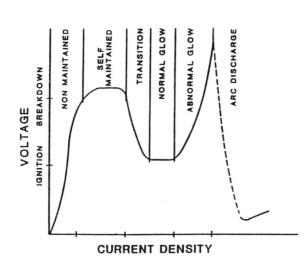

Fig. 4 Glow Discharge Electrical Characteristics

The geometry of this visible glow is important for uniform plasma nitriding. In those areas that are covered, the glow accurately replicates the physical shape of the product at a distance a few millimeters from its surface. Since the thickness of this visible glow is very small compared to the distance between the product surface and the vacuum retort, the operating point on the voltage/current density curve can be determined from conditions that exist very near to the product surface. The fact that the measured distance between the product surface and the vacuum retort is different at different locations in the load being plasma nitrided is not critical to the electrical characteristics of the normal glow discharge.

This region, where the visible glow can spread from a small local spot to cover the entire work surface, is called the Normal Glow Discharge Region. The voltage drop from the vacuum retort to the product surface remains nearly constant while the current density is increased to about 10^{-1} mA/cm^2 in the Normal Glow Discharge Region.

THE ABNORMAL GLOW

A further increase in current density brings the voltage/current density characteristic into the abnormal glow discharge region. In spite of this name, which has historical implications, this is the region useful for plasma nitriding. It is only in this abnormal glow discharge region that the product surface will be completely and uniformly covered by the visible glow. The current density, that is important for promoting the desired physical and chemical reactions of plasma nitriding, will be uniform and at the highest value that will still permit stable electrical control. Since the visible glow already covers all of the product surfaces, an increase in current density will now be accompanied by an increase in the voltage drop through the resistance of the glow discharge. This positive characteristic, where an increase in current density is accompanied by an increase in voltage drop, is desirable because it permits stable control of the operating point in the abnormal glow discharge region.

The current density in the Abnormal Glow Discharge Region is generally between .1 and 5.0 ma/cm^2 for voltage drops between 400 and 800 volts.

THE ARC DISCHARGE

If the power supply voltage is increased in the abnormal glow discharge region, the current density will increase. These increases in both the voltage and the current density will produce an increase in the power density and, therefore, the thermal energy delivered to the product surface through the plasma from the electrical power supply. If the delivered power is allowed to increase to values high enough to cause local overheating of the product surface, the resultant increase in electron emission will allow an additional increase in the current density. The glow discharge will concentrate itself in this overheated area and a high thermal energy arc discharge will occur. This arc discharge, if it is allowed to persist, will cause noticeable pitting and even melting of the product surface. Proper equipment design and good plasma nitriding practice must limit the frequency of arc formation.

These arc discharges can be turned off after a time interval that varies with the design quality of the power supply. Long switch-off times permit some damage to the product surface and switch-off times greater than 2 microseconds generally permit observable damage to polished surfaces. Fortunately, power supply software and hardware can interrupt an arc discharge, once formed, before a significant amount of power and thermal energy can be delivered to the product surface. In fact, arc discharge turn-off time is just one important indicator of power supply quality.

HOT WALL PLASMA NITRIDING EQUIPMENT

Equipment supplied for plasma nitriding in the abnormal glow discharge region generally resembles the conventional hot wall vacuum furnace shown in Figure 5.

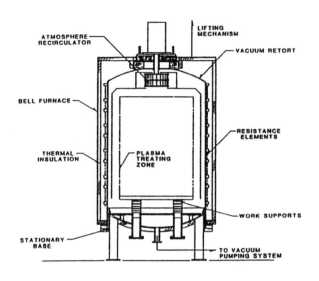

ATMOSPHERE RECIRCULATOR

LIFTING MECHANISM

VACUUM RETORT

BELL FURNACE

RESISTANCE ELEMENTS

THERMAL INSULATION

PLASMA TREATING ZONE

WORK SUPPORTS

STATIONARY BASE

TO VACUUM PUMPING SYSTEM

Fig. 5 Equipment for Plasma Heat Treating

The equipment will include a bell furnace heated with resistance elements and lined with lightweight thermal insulation to limit heat loss. A mechanism will be provided for lifting the bell furnace and the vacuum retort from the stationary base to provide complete access to the work supports for top or front loading. The base will contain electrically insulated hearth rails to support the work in the center of the plasma treating zone. A stationary base allows permanent utility connections, including connections to the process gas inlet and pumping system outlet lines. The fixtured work can remain stationary throughout the plasma treatment and can be instrumented with direct contact temperature sensors to monitor all heating and cooling cycles.

A vacuum pumping system will be supplied to initially purge the retort and then to maintain the partial pressures needed for plasma operation in the abnormal glow discharge region. The retort can be backfilled with an inert or protective atmosphere after vacuum purging to reduce the time for heating heavy loads to the operating temperature. An internal atmosphere recirculator can then be employed for heating the product inside the vacuum retort by gas convection . With this insulated, hot wall arrangement the retort wall can be efficiently operated at all temperatures at full atmospheric pressure.

The bell furnace can be removed for fast cooling without product surface damage, since the independent vacuum retort can contain the product within the recirculating protective atmosphere. The selection of lightweight fibrous insulation permits lifting and moving the bell furnace without spalling damage from rapid changes in operating temperature.

SOME OPERATING LIMITATIONS
It is important to consider the options available to the furnace operator when the voltage/current density is near the transition between the normal glow and the abnormal glow discharge regions shown in Figure 4. His reprogramming procedure must not alter the partial pressure or the gas mixture. Changes in these parameters will directly effect the chemistry and metallurgy at the product surface.

If the temperature is above the maximum value allowed by material considerations he must be able to lower the thermal input from the plasma without decreasing the voltage and the current density. These parameter changes would move the process into the unacceptable normal glow discharge region. If the voltage/current density is near the transition between the abnormal glow and the arc discharge regions and the temperature is below the minimum value required for proper thermal diffusion, he must be able to increase the thermal energy input without increasing the voltage and the current density. These parameter changes would move the process into the unacceptable arc discharge region. These operating limitations are easily managed when a pulsed plasma power supply with a variable duty cycle is employed.

POWER SUPPLY CHARACTERISTICS

Separate power supplies can be provided for heating the bell furnace and for maintaining the plasma voltage and current parameters. The bell furnace heaters can be operated from conventional SCR power supplies using standard PID controllers to maintain the desired nitriding temperature.

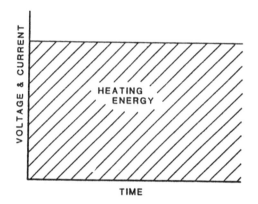

Fig. 6 Voltage and Current/Time With a DC Power Supply

Plasma power supplies require special features.[3] Both Continuous DC and Pulsed Power Supplies have been supplied for establishing the plasma parameters. With a DC power supply, a continuous voltage is applied to the system and, therefore, the plasma current density is also continuous. The power density at the product surface, that determines the amount of heating energy transferred through the plasma and the resulting product temperature, will be proportional to the area under the Voltage and Current/Time curve as shown in Figure 6. In this case, an increase or a decrease in the plasma voltage, to produce some desired variation in the positive nitrogen ion density, must be accompanied by an increase or decrease in the heating energy transferred from the plasma, whether this additional variation was desired or not.

The Metals Handbook notes this limitation.[4] It is first recognized that a hot wall furnace is desirable because of its energy efficiency, requiring only about 20% of the power that would be used by a similar cold wall furnace at $1060^{\circ}F$. But it is also noted that a DC power supply, operating without control of the power duty cycle, can overheat the product in a hot wall furnace at high current densities and that (in this case) adequate surface activation of high alloy and stainless steels may not be possible.

PLASMA DUTY CYCLE

The heating effect of the plasma can be limited in a system that permits regulation of the time the plasma is on and the time the plasma is off. Pulsed plasma power supplies, with a variable timed separation between pulses, permit the selection of a variable power duty cycle that can be defined as

power duty cycle = (time on)/(time on + time off).

With a pulsed power supply, the plasma voltage can be applied continuously for a 100% power duty cycle, or with equal on/off timed pulses, a 50% duty cycle.

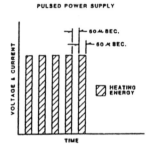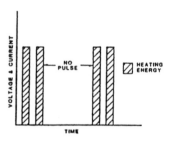

Voltage and Current/Time with a 50% Duty Cycle   Fig. 7   Voltage and Current/Time with a 5% Duty Cycle

Figure 7 shows the voltage and current relationship with time with a 10 kilohertz pulse frequency producing a 50% power duty cycle. This duty cycle can be reduced to 5% by extending the off-time interval without any undesirable effect on the positive nitrogen ion density at the product surface, also shown in Figure 7. Since the power density remains the area under the Voltage and Current/Time curve, the energy input through the plasma will now be so small the product will not be significantly heated by plasma energy. Independently controlled bell furnace resistance heaters can now be employed to bring the product to the proper temperature
for surface treatment. In normal operation, with standard programming practice, the current density and therefore the power density will not be a function of this power duty cycle. The current density can remain essentially independent of the power duty cycle as long as the power from the plasma is not allowed to heat the gas and the product remains completely covered by the glow discharge. Figure 8 shows the continuing direct relationship of the current density and voltage that remains nearly independent of the power duty cycle for changes in the duty cycle between 5% and 50% with an abnormal glow discharge.

OPERATING PARAMETERS FOR PLASMA NITRIDING

In the most ideal situation it will be possible for the furnace operator to select a

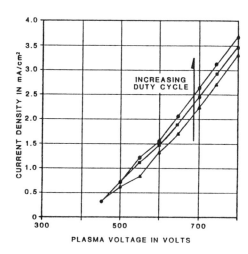

Fig. 8 Current Density =
f(Plasma Voltage & Duty Cycle)
For a Fixed Gas Pressure,
Surface Temperature and Gas Composition

plasma voltage to produce a short and efficient nitriding cycle, a gas mixture to produce the desired chemistry in the compound zone, a time duration to insure the specified depth of diffusion, and a surface temperature that will permit retaining the original core properties. The microprocessor will then automatically adjust these initial settings and the related dependent parameters to maintain process control.

In the worst situation it will be necessary for the furnace operator to personally evaluate the influence of the total product surface area on the final thermal energy balance that will be responsible for establishing the nitriding temperature. He will be forced to consider the relationship of the physical position of products within the load and their influence on the expected temperature uniformity. Periodic operator intervention will be required to inspect the load through the view port and to make setpoint adjustments based on intuitive knowledge, empirical data, optical pyrometer readings or the color and overall appearance of the glow discharge.

Ideally there are only a limited number of power supply, gas mixture, and product related parameters that must be independently selected by the process programmer for a plasma nitriding cycle to be run with microprocessor control.

     1. Pulse Voltage Amplitude

     2. Gas Mixture Partial Pressures

     3. Nitrogen, Hydrogen, and Methane Mass Flow Rates

     4. Pulse Time Duration and Repetition Rate

     5. Bell Furnace Zone Temperatures

     6. Nitriding Time to The Endpoint

The setpoints selected for the Pulse Voltage Amplitude, the Gas Mixture Partial Pressures, and the Nitrogen, Hydrogen, and Methane Mass Flow Rates will interact through the mechanism of the abnormal glow discharge to establish the activity of nitrogen at the surface of the product. Considerable operating experience indicates that, at least, this nitrogen activity and the product surface temperature must be independently set and controlled for a plasma nitriding process to be truly manageable at an industrial level by a furnace operator. Unnecessary interdependence between these parameters can make the specification of a plasma nitriding program extremely complex and often a matter requiring empirical determinations.

As an example, it is often necessary to vary the gas composition during the course of a single run to produce the desired surface layers. However, the voltage/current relationship, and therefore the thermal input from the glow discharge, depends in large measure upon this same gas composition. It is therefore necessary to separate the heating effect from the requirements for surface chemistry so that a desired change in gas composition is not inadvertently accompanied by an undesirable change in the temperature of the product surface. This operating requirement can be met with pulsed power supplies.

The pulse time duration and repetition rates initially selected by the furnace operator can limit the thermal energy transfer from the plasma to values that will not appreciably influence product surface temperature. Product temperature can be independently maintained by the bell furnace. The surface temperature can then be continuously monitored with direct contact thermocouples to ensure this limited thermal energy transfer through the plasma. If the parameters selected for nitrogen activity conspire to increase the product surface temperature, the microprocessor software can adjust the pulse duration and the pulse repetition rate to lower the thermal energy transfer to an acceptable value. Under these conditions the bell furnace zone temperatures will control product temperature and, therefore, the rate of nitrogen diffusion from the surface of the product to the core. The Bell Furnace Temperature and the Nitriding Time to Endpoint will help determine the final nitride metallurgy.

PLASMA NITRIDING ADVANTAGES

Plasma nitriding differs from other nitriding technologies by using the phenomena of the electrical glow discharge to activate the nitrogen gas molecules needed for the process. There are important consequences to activating nitrogen in this manner. Plasma nitriding can be at temperatures below those employed for conventional nitriding, retaining maximum core properties. Sputtering at low temperatures to initially remove superficial contamination from the product surface permits short nitriding times even at these lower temperatures. Plasma nitriding permits better control over the composition of the final product surface, its structure, and its properties. It is an effective surface treatment for non-ferrous materials as well as cast iron and alloyed steels.

Plasma nitriding is environmentally non polluting. The gas discharge from a plasma nitriding furnace is non-toxic and non-explosive and can be vented directly to an outdoor location. It eliminates expensive product surface post cleaning operations and the final disposal of the cleaning solutions themselves in an EPA approved manner. Companies forced to retire their liquid nitriding equipment, anticipating even more stringent regulation, often select the most advanced alternative technology if there are no overriding commercial reasons for not doing so.

Although plasma systems may not compete in every situation for this reason alone, plasma systems are now being installed to satisfy environmental requirements.

Plasma systems can improve the uniformity of the hardness profile on complex sections. Aluminum extrusion dies and hot work forging dies perform better after plasma nitriding because all of the critical bearing surfaces can be reached by the nitriding plasma. Plasma nitriding systems are uniquely qualified to penetrate openings with small dimensions to supply active nitrogen neutrals and ions to surfaces that might otherwise not be hardened because of a deficient atmosphere recirculation system.[5]

And finally, plasma nitriding permits the development of superficial hardness on sintered parts with thin sections without concern that certain of these parts may be made brittle by unexpected through hardening. The plasma nitriding operation does not require the added cost and complexity of stop-off infiltration, copper addition to the powder mix or steam blueing of the sintered part to prevent this undesirable through hardening. It is the finite thickness of the cathode fall region that prevents undesirable nitride penetration of parts that have been compacted and sintered at less than full density.[6]

1. Grube, W. L. (1978), "High Rate Carburizing in a Glow Discharge Methane Plasma," Metallurgical Trans. A AIME, 9A pp 1421-1429.

2. Sudarshan, T. S. (1989),"Surface Modification Technologies - An Engineer's Guide," Marcel Dekker, Inc., pp 323-328.

3. Straemke, S. W. (1986), "Theoretical Process Considerations," ASM Internationals 1st Ionitriding Conference, Cleveland, OH.

4. Metals Handbook, 9th ed., Vol. 4, Heat Treating, American Society for Metals, pp 213-216.

5. Dressler, S.(1988), "Plasma Nitriding Extends Die Life," Proceedings Of The Fourth Int'l Aluminum Extrusion Technology Seminar, Chicago, IL, Vol - 2 pp 103-108.

6. Chen, Y. T., "Surface Treatment Of P/M Steels By Ion Nitriding," Pitney Bowes Inc., Stamford, CT.

IONIC NITRIDING AND IONIC CARBURIZING
OF PURE TITANIUM AND ITS ALLOYS

Jean-Pierre Souchard
Société B. M. I (Groupe H.I.T.)
St. Quentin Fallavier
38290 La Verpilliere, France

Patrick Jacquot, Bernard Coll, Marc Buvron
Innovatique S. A. (Groupe H.I.T.)
Place Charles Andrieu
60530 Neuilly-En-Thelle, France

ABSTRACT

The surface hardness, the wear and corrosion resistance of titanium and titanium alloy can be improved by thermochemical surface treatment.
Plasma heat treatment has many advantages over other surface treatment methods because of the great number of independent process parameters which enables layers with specific microstructures, hardness and properties to be produced. It is the reason why we carried out a study to verify the role of different treatment parameters. The plasma heat treatment was performed with an industrial plasma furnace equiped with a graphite resistor to minimize the plasma power and to heat the specimens. The plasma source is given by a pulsed power supply.
The results of the investigations have shown the effect of different process parameters onto the metallurgical characteristics of plasma nitrided and carburized titanium alloy. It was shown that plasma heat treatment of titanium alloys, in given conditions, resulted in the formation of titanium nitride or titanium carbide layers.
After plasma nitriding of 6 hours of pure titanium or Ti6A14V alloy at 900 °C, the surface hardness can be increased of more five time the core-hardness of the base material.

IT IS WELL KNOWN THAT titanium alloys have inherent advantages of light weight and low modulus. However, their tribological behaviours are characterised by a high coefficient of friction and poor wear resistance. By using plasma nitriding or ion carburizing processes, it is possible to enhance the tribological properties of titanium alloys.

Recently, the interest for improving the wear resistance of titanium alloys has been renewed by the use of the plasma nitriding technique (1) (2) (3) (4). It seems that titanium nitride or carbide layers with high hardnesses and low coefficient of friction can be carried out.

In this work, we have looked for the influence of different process parameters, such as : temperature, time, pressure and electric parameters on the microstructure, microhardness, thickness of compound layer, depth of diffusion layer, grain size and roughness.

The purpose of this paper is to present the results of these investigations.

EXPERIMENTAL PROCEDURE

Specimens for treatment (30 X 20 X 20 mm) were polished and degreased with aceton at room temperature. The composition of the materials used in this study are : pure titanium (annealed, alpha alloy), alloy Ti6A14V (annealed, alpha-beta alloy).

The plasma surface treatment was performed in a small industrial furnace (B.M.I. - VB 50 X 50), which is shown shematically in Fig. 1. This system differs from the conventional nitriding process in two points. Firstly, the cathodic holder is made of titanium. Hence the possibility of contamination by sputtering of some foreign materials onto the parts is eliminated. Secondly, the parts are heated by an additional heating system. It consists of 12 graphite resistance elements, completely surrounding the load to give uniform heating, and connected to a 45kW power supply. The hot zone is thermally insulated with graphite fibers insulator.

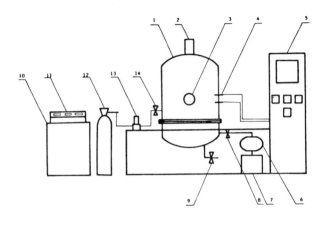

1	Vessel	8	Control valve
2	Turbine	9	Venting valve
3	View port	10	Power unit
4	Thermocouple	11	Power control panel
5	Control panel	12	Gas
6	Roots pump	13	Mass flowmeters
7	Rotary pump	14	Gas control valve

Fig. 1 : Schematic representation of plasma heat treatment equipment

The useful volume of the vacuum chamber is about 80 liters (∅ 450 X500 mm high).

The specimens were connected to the cathodic bias of a pulsed power supply (20 kW/ 6-7KHz) and the heating elements serve as the anode.

During the plasma treatment, the voltage was set between 200 and 400 volts and current about 10 Amp. Temperature was controlled using an infrared pyrometer placed in front of a quartz view port. Nitriding and carburizing were carried out at a working pressure of 6 mbar with a constant gas flowrate of 1,4 l/mn. The gas composition was respectively either pure nitrogen or propane hydrogen mixture

The specimens were etched during 10 mn by an argon glow discharge. After this stage, the parts were heated with the heating elements in a pure nitrogen atmosphere until the treatment temperature was reached.

The plasma nitriding of titanium alloy

was performed at four different temperatures between 750 °C and 900 °C. Furthermore, for each temperature, the nitriding cycle time was established as following : 1, 3, 6 and 9 h.

The plasma carburizing of pure titanium was performed at 800 °C, 850 °C and 900 °C during 6 h.

After treatment, the specimens were polished, etched using a mixture of HF and HNO_3 acids and examined by optical metallographic technique. Microhardness profiles were used to determine the diffusion layer depth (depth = Hv0,05 core hardness + 100 HV).

IONIC NITRIDING OF PURE TITANIUM AND TITANIUM ALLOY

INFLUENCE OF THE TEMPERATURE AND THE TREATMENT TIME

The compound and diffusion layers – We have studied the influence of different temperature and treatment time on the compound layer TiN + Ti_2N, achieved after nitriding of pure titanium and titanium alloy (Fig. 2).

The compound layer thickness are given in Table 1, in function of nitriding temperature and time.

Temperature (°C)	750		800		850		900	
Time (h)	Ti6Al4V	Ti	Ti6Al4V	Ti	Ti6Al4V	Ti	Ti6Al4V	Ti
1	0,5	0,5	1	1	1	1,25	1	2
3	1	1,5	1,5	1,5	2	2	2	2
6	1,5	2	2	2	2	2	2	2
9	2	2	3	3	4	4	6	6

Table 1 : Ionic nitriding of pure titanium and Ti6Al4V alloy. Influence of time and temperature on compound layer thickness (in µm $\pm$ 0,5)

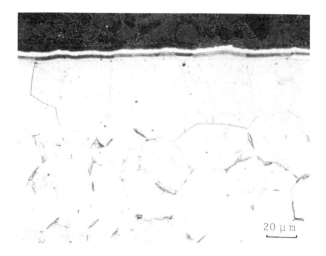

Fig. 2 : Optical micrograph of cross section
of titanium plasma nitrided in pure
nitrogen for 9 h at 800 °C.

We can see that the thickness of the compound layer, composed of the nitrides TiN and
Ti_2N, increases with nitriding temperature and
time. If the nitriding time is less than 9 h,
the thickness of the compound layer is found
to be inferior or equal to 2 µm for both materials.

Between 800 and 900°C, and for a treatment time of 9 h, the compound thickness
increased rapidly up to 3 - 6 µm. The variation with temperature and time of diffusion
layer on pure titanium and Ti6Al4V alloy
plasma nitrided are given in Fig. 3 and 4.
The thickness of the diffusion layer is mesu-
red with microhardness profiles using the
core hardness + 100 Hv.

Diffusion layer
thickness (mm)

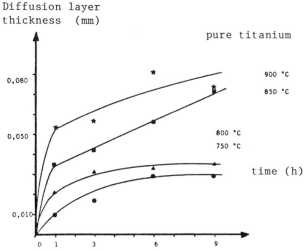

Fig. 3 : Ionic nitriding of pure titanium.
Variation with temperature and time
of the diffusion layer thickness.

Diffusion layer
thickness (mm)

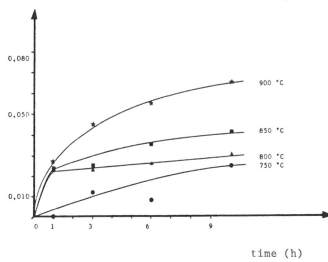

Fig. 4 : Ionic nitriding of Ti6Al4V alloy.
Variation with temperature and time
of the diffusion layer thickness.

The curves indicates that :
- the growth of both layers is controlled
  essentially by a diffusion mechanism,
- the maximal thickness of the compound layer
  (6 µm) is obtained for a treatment at 900°C
  during 9 hours (Fig. 5),
- concerning the thickness of compound layer,
  there is no significant difference between
  pure titanium and Ti6Al4V alloy.
Sometimes, the compound layer of
Ti6Al4V is a little bit thinner than that a
pure titanium.

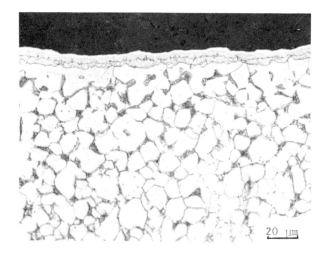

Fig. 5 : Optical micrograph of cross section
of Ti6Al4V alloy plasma nitrided in
pure nitrogen for 9 h at 900 °C.

- for the same nitriding conditions, the depth of the diffusion layer is always thicker on pure titanium than on Ti6A14V. Perhaps, in the presence of alloy elements, such as aluminium and vanadium, the Ti6A14V nitrided surface acts as a diffusion barrier (TiN, Ti_2N, Ti_2AlN) this retarding diffusion layer formation (5).

- the depth of the diffusion layer increases slowly with treatment time, but quickly with nitriding temperature.

Superficial microhardness
HV 0,05

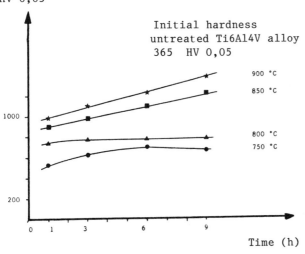

Fig. 6 : Ionic nitriding of Ti6A14V alloy. Superficial microhardness (HV0,05 ± 5 %) as a function of temperature and time.

Superficial microhardness
HV 0,05

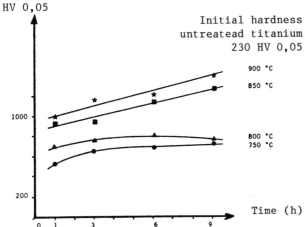

Fig. 7 : Ionic nitriding of pure titanium. Superficial microhardness (HV0,05 ± 5 %) as a function of temperature and time.

The superficial microhardness - Micro-hardness of the nitrided surfaces of specimens were measured using a Vickers indenter with a 50 g load.

The influence of temperature and nitriding time on the superficial hardness is shown on figures 6 and 7.

The superficial hardness increases conti-nuiously with time and particularly with the temperature. Similar observations were made by METIN and INAL (6). These results show that the superficial hardness level depends on the TiN +Ti_2N layer thickness and the hardness of the diffusion layer.

The grain size - We have investigated the effect of temperature and time on grain size of Ti and Ti6A14V samples. The average size of the grains increases with temperature and time of the treatment.

The grain size variations are not very important for Ti6A14V alloy, except for the treatment at 850°C/900°C during 9 hours.

In the case of pure titanium, long treat-ment time at 900°C results in an important and irreversible grain growth.

The surface roughness - The surface roughness is deteriorated when the temperature and time increase (Fig. 8).

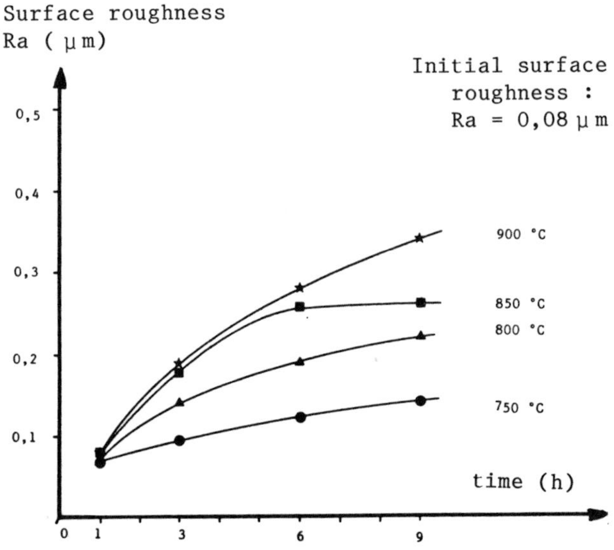

Surface roughness
Ra (μ m)

Fig. 8 : Surface roughness as a function of temperature and time - material : Ti6A14V alloy

For a nitriding treatment at 900°C during 9 hours, the surface roughness is 450 % up on the initial value (0,08 to 0,35 μm).

This can be explained by the bombardment effect occuring during the ionic etching.

INFLUENCE OF THE PRESSURE - Different parameters have been tested to determine the influence of operating pressure.

Firstly, we have examined the compound thickness after plasma nitriding at 850 °C during 6 hours for 3 pressure levels : 2, 4 and 6 mbar.

The results presented in Table 2 show that the thicker compound layer is obtained when the pressure decreases at about 2 mbar.

It seems that the sputtering effect occuring with the glow discharge can perhaps explained these results.

Pressure (mbar)	Surface roughness Ra (μm)
2	0,22
4	0,25
6	0,26

Initial roughness = 0,08 μm

Table 3 : Ionic nitriding of titanium at 850°C during 6 h. Influence of the pressure on the surface roughness.

INFLUENCE OF ELECTRICAL PAREMETERS - As previously stated, all treatments were carried out using a pulsed plasma (6-7 KHz) obtained by switching the discharge power supply on and off at microsecond intervals.

For different plasma nitriding cycles, we have studied the influence of the electrical impulse duration, corresponding to the period fraction, called : cyclic ratio or Rc (Fig. 9).

	Compound thickness ($\pm$ 0,5 μm)	
Substrate Pressure	Titanium	Ti6A14V
2 mbar	3 μm	3 μm
4 mbar	2 μm	2 μm
6 mbar	2 μm	2 μm

Table 2 : Ionic nitriding of pure titanium and Ti6A14V alloy at 850°C during 6 h. Influence of operating pressure on the compound thickness.

Secondly we have examined the influence of the pressure on the diffusion layer depth and the superficial microhardness. No significant difference was noticed. Finally we have measured the surface roughness for the 3 different pressure levels. The surface roughness increases when the pressure increases from 2 mbar to 6 mbar (see Table 3).

Intensity (A)

Cyclic ratio : Rc = $\dfrac{a}{t}$

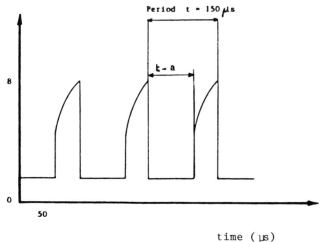

Fig. 9 : Shape of the current on an oscilloscope

Experiments were carried out to determine the influence of different cyclic ratio "Rc (1/5, 1/4, 1/3) on the compound and diffusion layers.

It appears that these both layers thickness decrease very slowly when "Rc" decreases from value 1/3 to 1/5. This result is not very significant.

On the other hand, a surprising influence of the cyclic ratio on the superficial microhardness of the ion nitrided titanium and Ti6A14V alloy was observed. For instance, with increasing "Rc" from 1/5 to 1/3, the superficial microhardness of the ion nitrided pure titanium increases from about 900 to more 1100 HV 0,005.

The influence of Rc on the surface roughness was investigated. The results are shown in Table 4.

It appears that the surface finish is deteriorated when the cyclic ratio increases. These results are probably related to the ion bombardment effect, that is more intense when the cyclic ratio or the electric impulse duration increases.

Cyclic ratio Rc	Surface roughness Ra (µm)
1/5	0,08
1/4	0,16
1/3	0,26

Initial roughness = 0,08 µm

Table 4 : Ionic nitriding of titanium and Ti6A14V alloy at 850 °C during 6 h, pressure : 6 mbar, intensity : 8 A Influence of the cyclic ratio on the surface roughness

IONIC CARBURIZING OF PURE TITANIUM

Ionic carburizing treatment was performed in a pulsed plasma of a propane and hydrogen gas mixture.

In order to show the effects of gas carbon content on the microstructure, thickness and microhardness profile of the carburized layers, the plasma heat treatment was carried out in gas mixtures containing 3, 10 and 20 at % C in hydrogen.

After etching with an hydrogen glow discharge, the parts are heated in the same gas with the additional heating system. The specimens are all treated at 900 °C for 6 hours and the gas pressure was adjusted to 6 mbar.

Microscopic examination has revealed that the compound layer produced with different carbon content in the atmosphere had an uniform and homogeneous microstructure.

After treatment, in the gas mixture with 10 and 20 at % carbon, the specimens are formed :
- immediatly on the surface, by a dense and amorphous carbon film (Fig. 10),
- by a very thin and almost pore-free TiC (Titanium Carbide) compound layer,
- and by a diffusion layer.

Fig. 10 : Optical micrograph of cross section of pure titanium plasma carburized in 20 at % C ($C_3H_8-H_2$) for 6 h at 900 °C.

The thickness of these different layers is shown in Table 5.

at % C in $C_3H_8-H_2$	Thickness (microns)	
	TiC	a – C
3	4	0
10	1,5	3
20	1	6

Table 5 : Thickness of the TiC and amorphous carbon layers versus % C in the $C_3H_8-H_2$ gas mixture obtained after carburizing at 900°C during 6 h.

The use of 3 at % carbon in the atmosphere formed a TiC compound layer of 4 μm thick and suppressed the formation of the amorphous carbon film (Fig. 11).

Fig. 12 shows the hardness profiles of ion carburized pure titanium with 3 different gas carbon contents. We can observe that the case depth level is higher for decreasing carbon content in the gas atmosphere.

The results we have obtained show that when the gas atmosphere contain more that 3 at % C, the rate of carbon mass transfer to the sample surface exceed the rate of diffusion into the sample. It is the reason why when the treatment gas is too rich in carbon an amorphous carbon film is formed on the surface and hindered the carbon migration into the diffusion layer.

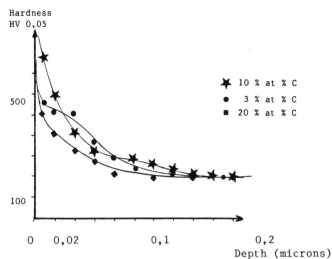

Fig. 12 : Ionic carburizing of pure titanium at 900 °C during 6 h. Microhardness profiles evolution versus % C in the $C_3H_8-H_2$ gas mixture.

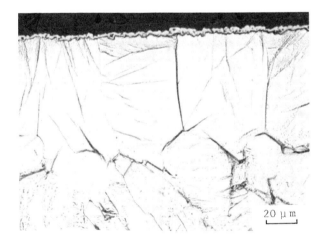

Fig. 11 : Optical micrograph of cross section of pure titanium plasma carburized in 3 at % C ($C_3H_8-H_2$) for 6 h at 900 °C.

CONCLUSION

The effect of temperature and nitriding time on different physical parameters of plasma nitrided titanium and titanium alloy Ti6Al4V has been investigated.

The results obtained can be summarized as follows : a rise in temperature and treatment time result in an increase of :

- the compound layer thickness,
- the diffusion layer depth,
- the superficial microhardness,
- the surface roughness,
- the grain size.

A low operating pressure (2 mbar) seems to be a suitable factor to obtain the maximal compound layer thickness and a smooth surface finish.

The influence of the electrical impulse duration used in pulsed plasma nitriding was investigated. It was found that the best surface finish is obtained for the lawer cyclic ratio value and on contrary, the surface microhardness is maximum for a highest cyclic ratio value.

Plasma carburizing of pure titanium in a gas containing 3 at % C results in the formation of titanium carbide in the compound layer and a carbon diffusion layer in the titanium matrix.

At higher gas mixture carbon contents, compound layer formation is slowed down by the formation of an amorphous carbon thin film.

REFERENCES

(1) E. Rolinski
     Surface Eng. (1986), Vol. 2, N° 1, p.35-42

(2) K.T. Rie, Th. Lampe, S. Eisenberg
     Surface Eng. (1985), Vol. 1, N° 3,
     p. 198-202

(3) A. Raveh, G. Kimmel, U. Carmi,
     A. Inspektor, A. Grill, R. Avni
     Surface and Coatings Technology,
     36 (1988), p. 183-190

(4) J.M. Molarius and al
     Härterei - Techn. Mitt. 41 (1986) 6,
     p. 391-397

(5) Th. Lampe, S. Eisenberg
     Intern. Seminar on Plasma Heat Treatment
     Senlis (France) 21-23 Sept 1987
     - PYC Edition

(6) E. Metin, O.T. Inal
     Proceedings Intern. Conf. on Ion Nitriding
     ASM, Cleveland (U.S.A.), 15-17 Sept 1989

EFFECT OF GEOMETRY ON GROWTH
OF NITRIDE LAYER IN ION-NITRIDING

M. J. Park
University of Hull
United Kingdom

W. S. Baek, S. C. Kwon
Korea Institute of Machinery and Metals
Changwon, Kyungnam, Republic of Korea

M. C. Yoo
University of Illinois
Urbana-Champaign, USA

Abstract

The growth behavior of nitride layer inside a long hollow tube with intricate geometrical factor was investigated in order to find out a way to enhance he uniformity of nitride layer. The inner surface of steel tube of 30mm inner diameter was machined to have corrugation depth ranging from 0.65mm to 3.90mm and corrugation width from 1.10mm to 13.0mm and was put into ion-nitriding at 525 °C for 10 hours in 2.5 torr operating pressure. The thickness of compound layer and diffusion layer on the land and the groove was measured and analyzed according to corrugation depth and width. As the corrugation became deeper, the growth of the compound layer on the land increased and the growth of the layer on the groove decreased. the thickness variation of diffusion layer on the land and the groove despicted similar tendency to that of the compound layer. As the corrugation became wider, the growth of compound layer on the land decreased and the growth of the layer on the groove increased. The thickness variation of the diffusion layer on the land and the groove resembled that of of the compound layer.

THE SURFACE HARDENING OF MATERIALS has been a persistant problems to materials scientists and engineers. One of the popular processes to achieve hardening of steel surface is the nitriding which has been in commercial use since early 20th century(1). The nitriding process saves the subsequent quenching because of its comparatively low processing temperature of 450 °C to 570 °C, and the distortion and dimmensional changes of workpieces is lessened with nitriding than with carburizing(2). The conventional nitriding is customarily carried out in an atmosphere of partially dissociated ammonia, or in a cyanide-cyanate salt bath which includes potential environmental hazards and difficulties in controlling the growth of white layer.

The newer nitriding process was patented in 1931(2) and has been getting wide attention as an industrial alternative to the conventional gas or salt bath nitriding of steel(4). The new process, so-called, ion-nitriding is accomplished by using the energy of the glow discharge. This ion-nitriding process is characterized by its shorter process duration, cleanliness and in particular its complete non-toxity as well as is easy to control the growth of the undesirable white layer(5,6).

Most studies to date has been concerned with the evaluation of engineering valuables such as the alloying elements, temperature and time duration of ion-nitriding along with the gas mixtures and the operating pressure (7,8,9). The ion-nitriding has been compared with the various nitriding processes from the stand point of its economic viability(10). Further development of the ion nitriding was made to suppress the arcing by supplying the pulsed direct current instead of direct current (11,12). The ion-nitriding characteristics with the pulsed glow discharge was studied in the academic view point by authors (13).

Production engineering demands for low production costs in addition to increase the wear resistance and fatique strength of workpieces by the surface hardening. No systematic work was undertaken to investigate the nitride layer growth inside the long cylinder with intricate geometrical shage. The present objective is to examine the growth behavior of nitride layer over the corrugated internal surfaces in a long hollow tube and to find out enengineering design factors to enhance the uniformity of nitrided layer.

EXPERIMENTAL PROCEDURE

A commercial grade of Cr-Mo steel, whose chemical composition is given in Table 1, was used in this investigation.

Specimens were bored and machined to hollow cylinder of 200(L) x 30(ID) x 50(OD) mm size. The internal surface of the hollow cylinder was machined to make a corrugation as shown in Fig. 1.

Table 1 - Chemical Composition of Specimen
(wt.%)

C	Si	Mn	P	S	Cr	Mo	Fe
0.43	0.24	0.67	0.026	0.024	0.98	0.24	Bal.

The corrugation depth of specimen was varied from 0.65mm to 3.90mm and the corrugation width of specimen was changed from 1.10mm to 13.0mm. Table. 2 Summarzed the corrugation depth and width of the speciemen

Table 2 - Corrugation Depth and Width of the Specimen.

Position Specimen	Depth (mm) (d)	Width (mm) (w)
1	0.65	2.35
2	1.30	2.35
3	2.60	2.35
4	3.90	2.35
5	0.65	1.10
6	0.65	4.60
7	0.65	6.90
8	0.65	13.0

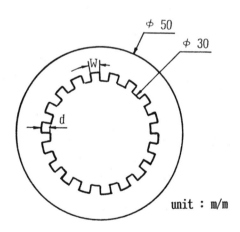

Fig. 1 Schematic Drawing of Corrugation

Inside Hollow Cylinder

The specimens were thoroughly degreased, descaled, washed, dried, and positioned on the cathode stage in the ion-nitriding chamber as in Fig. 2. The anode was a stainless steel rod of 6 mm in diameter and positioned along the hollow cylinder in side the specimen. Before ion-nitriding the specimen, a mixture of nitrogen and hydrogen gas (1:1) was used after evacuation to clean the surface by sputter cleaning. The gas mixing ratio of nitrogen to hydrogen was converted from 1 to 4 and the operating pressure was maintained at o.4 torr throughout. The pulsed power of 20KW was employed in this work and the voltage of discharge was 500V. The specimens were nitrided at 525 °C for 10 hours. The temperature of specimen was measured with Chromel - Alumel thermocouple embedded in cathode stage. The glow discharge power supply was cut off at the end of treatment time and the specimen was left to cool in vacuum.

After ion-nitriding, the specimens were taken out and sectioned circumferentially with cutting wheel at the point of 10, 50, 100, 150, 190 mm from the bottom. Each sectioned pieces were mounted, metallographically prepared and etched with 3% Nital. The compound layer thickness was determined with an optical microscope equipped with a micrometer. The diffusion layer thickness was determined from the surface to the point where hardness value is 10% higher than the substracte from the hardness profile curve.

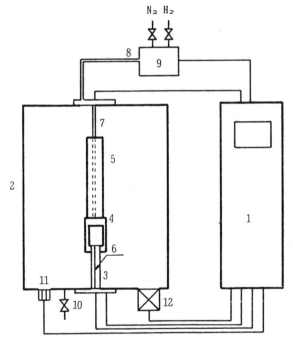

Fig. 2 Schematic Diagram of Pulse Ion-nitriding System

1.Control part and power supply 2.Chamber(anode) :ground 3.Cathode 4.Specimen holder 5.Specimen 6.Thermocouple 7.Auxiliary anode 8.Gas inlet 9.Gas control part 10.Leak valve 11.Pressure sensor 12.Rotary pump

RESULTS AND DISCUSSION

THE THICKNESS VARIATION OF NITRIDE COMPOUND LAYER OVER THE CORRUGATED SURFACE - Fig. 3 is a typical microstrcuture of nitrided layer on the land and the groove of the corrugation. The nitrided layer is thicker on the land than the groove. When the geometrical factor was considered on the thickness of nitride layer, the thickness reduced in order from the edge of land, the land, the groove to the edge of groove. The compound thickness variation over the corrugation was known to depend upon the HCD effect (6), nitrogen concentration (10), distribution of electric field (8) and the probability of compound adsorption (10). The above four factors except the distribution of electric field satisfied to explain the nitride thickness distribution over the corrugation inside the hollow cylinder. It may be said that the effect of electric field on the nitride layer growth is less dominant than the resultant effect of HCD, nitrogen concentration and probability of compound adsorption.

EFFECT OF CORRUGATION DEPTH- Fig.4 is the microstructure of the land and the groove at the middle point of the specimen with various corrugation depth. The nitride compound layer thickness on the land and the groove were measured and summarized in Table 3.

Table 3. Thickness of Nitride Compound on Land and Groove of Specimen with Various Corrugation Depth

Corrugation Depth (mm)	Thickness of Nitride Compound(μm)	
	Land	Groove
0.65	21	14
1.3	23	13
2.6	30	12
3.9	32	10

As the depth of corrugation increased,the thickness of the compound on the land increased from $21\,\mu$m to $32\,\mu$m and the layer thickness on the groove decreased from 14 to $10\,\mu$m. The variation of the compound layer thickness on the land and the groove was despicted in Fig. 5. The thicker growth of nitride layer on the land than the groove was resulted from the fact that the groove faced with stronger HCD effect, lower nitrogen concentration, and lower probability of compound absorption than than the land. Fig. 6. was the diffusion layer thickness variation on the land and the groove according to the depth of corrugation. The similar growth growth pattern of diffusion layer to that of compound layer was appeared as was generally observed in former investigations (14).

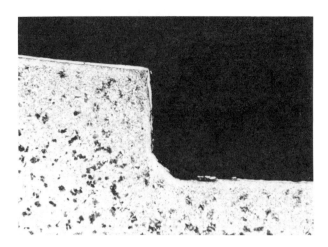

Fig. 3   Microstructure of Nitride Layer on Corrugation (x50)

EFFECT OF CORRUGATION WIDTH- Fig. 7 is the microstructure of the land and groove at the middle point of the specimen with various corrugation width. The nitride compound layer thickness on the land and the groove were measured and tabulated as in Table 4.

Table 4. Thickness of Nitride Compound on Land and Groove of Specimen with Various Corrugation Width

Corrugation Width (mm)	Thickness of Nitride Compound(μm)	
	Land	Groove
1.1	24	13
2.35	21	14
4.6	19	15
6.9	19	15
13.2	17	16

As the width of corrugation increased, the thickness of compound on the land decreased from $24\,\mu$m to $17\,\mu$m and the layer thickness on the groove increased from $13\,\mu$m to $16\,\mu$m. The compound layer thickness variation on the land and the groove was demonstrated in Fig. 8. while the diffusion layer thickness variation in Fig. 9. the nitride compound layer became thinner on the land and the layer on the groove thickner as the width of corrugation increased.These two curves converged converged to the compound layer thickness of $17\,\mu$m as the width of groove approached to over $13.2\,\mu$m. The probability of absorption of compound seemed to be dominant in the explanation the this phenomena. it was conlusive that both of the depth and the width of corrugation brought about geometrical complexity which differentiated severely the growth of compound and diffusion layer on the land and the groove respectively.

(a)
0.65mm

(b)
1.30mm

(c)
2.60mm

(d)
3.90mm

A) Land (x200)   B) Groove (x200)

Fig. 4   Microstructure of Nitride Layer on Land (A) and Groove (B)
          of Various   Corrugation Depth

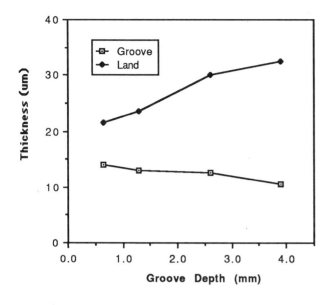

Fig. 5 Variation of Nitride Compound Layer Thickness
with Groove Depth

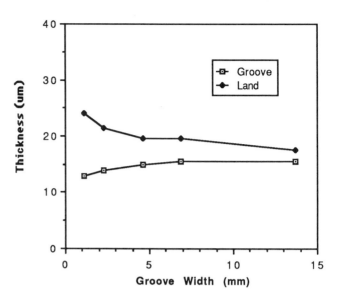

Fig. 8 Variation of Nitride Compound Layer Thickness
with Groove Width

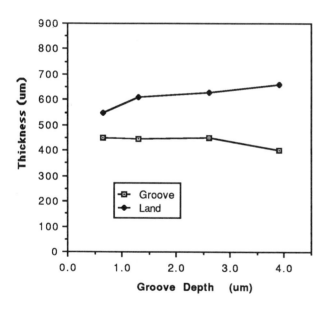

Fig. 6 Variation of Nitride Diffusion Layer Thickness
with Groove Depth

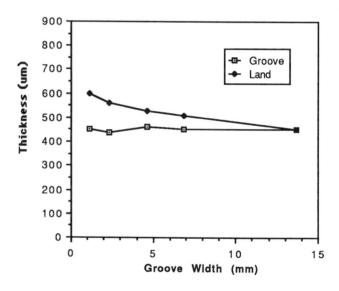

Fig. 9 Variation of Nitride Diffusion Layer Thickness
with Groove Width

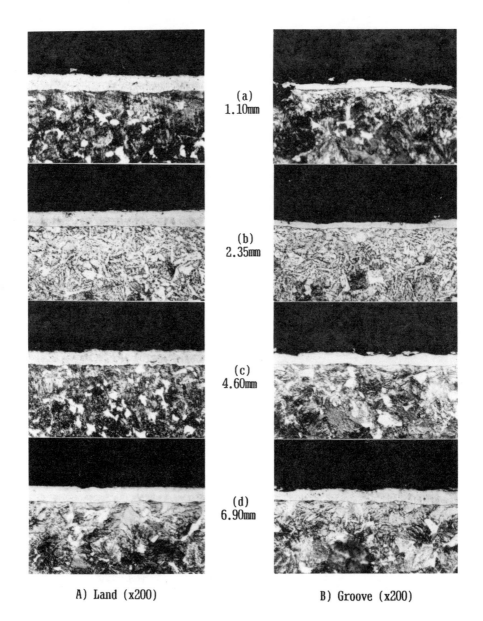

(a)
1.10mm

(b)
2.35mm

(c)
4.60mm

(d)
6.90mm

A) Land (x200)   B) Groove (x200)

Fig. 7   Microstructure of Nitride Layer on Land (A) and Groove (B)
          of Various Corrugation Width

CONCLUSIONS

The Cr-Mo steel hollow tube with corrugation over the internal surface was ion-nitrided for 10 hours at 525 °C in 2.5 torr operating pressure. The depth of corrugation was varied from 0.65mm to 3.90mm and the width of corrugation from 1.10mm to 13.0mm. The following conclusions were made on the growth characteristics of the nitride compound layer and the diffusion layer on the geometry.

(1) As the depth of the corrugation increased, the growth of the nitride compound layer on the land increased and the layer on the groove decreased. This phenomena was resulted from the HCD effect, the nitrogen concentration and the probability of compound absorption. The diffusion layer growth on the land and the groove despicted similar tendency to the compound layer growth on the corrugation.

(2) As the width of corrugation increased, the growth of the compound on the land decreased and the layer on the groove increased respectively. This was explained by the dominant effect of nitrogen concentration. The diffusion layer growth on the land and the groove showed similar to the nitride compound layer growth on the respective geometry.

(3) Both the depth and the width of corrugation represented the complexity of geometry which showed the close relationship with enhanced difference in the nitride layer growth on the land and on the groove.

ACKNOWLEDGEMENT

The authors are grateful to S.S Cha and I.K. Choi for providing technical assistance in the course of this research.

REFERENCES

1. A. Fry, Kruppshe Monatschefte, 43, 137-151 (1923)

2. A.M.Stainess and T.Bell, Thin Solid Film, 86, 201 (1981)

3. J.J.Egan, U.S.Patent 1837256,(1931)

4. B.Edenhofer, Metal Progress, 109, 38 (1976)

5. A.U.Seybolt, Trans. Metall. Soc. AIME, 245, 769 (1969)

6. V.A.Phillips and A.U.Seybolt, Trans. Metall. Soc. AIME, 242, 2415 (1968)

7. B.Edenhofer, Metall. Mater. Technol., 8, 421 (1976)

8. V.Korotchenko and T.Bell, Heat Treat.Met., 4,88 (1978)

9. C.K.Jones, S.W.Martin, D.J.Sturges,and M.Hudis, Heat Treatmont 1973, Metal Soc., London, 71-76 (1975)

10. F. Hombeck, Heat Treatment shanghai '83, Metal Soc, 41-49 (1984)

11. H.Wilhelmi, S.Straemke and H.C.Pohl, Harterei-Technische Mitteilungen, 37, 263 91982)

12. R.Gruen, International Conference on Ion Nitriding, Cleveland, ASM, 143-147 (1986)

13. S.C.Kwon, G.H.Lee and M.C.Yoo, ibid, 77-81 (1986)

14. P.C.Jindal, J.Vac. Sci. Technol., 15, 313-317 (1978)

CONVENTIONAL SCR FIRED PLASMA POWER SUPPLIES VS. PULSED HIGH FREQUENCY POWER SUPPLIES

William L. Kovacs
Elatec, Inc.
Andover, Massachusetts, 01810, USA

For the past 20 years the use of conventional "SCR fired" DC plasma power supplies have become accepted as reliable, efficient power supplies for conducting the ion nitriding process. In the past, the major controversy has been whether to use auxiliary heating with the DC power supply to supply supplemental heat to the load as well as power to dissociate the gas for the nitriding process. Under laboratory test conditions (with perfectly clean parts), the DC only heated system had an economic advantage, as it required only a single power supply. With perfectly clean parts, sputter cleaning time, and heating the parts to temperature were not an issue. Consequently, arguments were made that this was the more efficient and economic system for ion nitriding.

The problem arose in the real world when "dirty" or partially clean parts were used, or parts were used which had unusual geometries (or were improperly fixtured), causing hollow cathode discharge. Under these circumstances, a conventional SCR fired DC only heated plasma power supply required substantial operator skill to bring the load to temperature (if it could be brought to temperature at all), and often required sputter clean times ranging from two hours to 48 hours.

Additionally, the use of large production size SCR fired DC only heating power supplies also required extensive arc detection and suppression circuitry. Once the SCR had fired and an arc was detected, it was necessary to have circuitry which would, in effect, short circuit, or "crowbar" the power from the load to prevent arc damage. Arc detection circuits involved the following:

a) A maximum current limit (Imax).

b) Measurement of current increase which indicated start of arcing (di/dt).

c) Measurement of a voltage drop which was an indication that an arc had formed (dV/dt).

In these circumstances with use of DC only heated SCR fired power supplies, it was necessary to operate in the low current range over an initial critical part of the ion nitriding cycle requiring a large turndown ratio on the power supply. The single simple SCR fired DC only heated power supply became a costly, complex power supply which negated many of the economic and operational advantages claimed for heating by large DC only power.

The use of auxiliary heat eliminated the turndown problem as the DC power supplies required were much smaller. The typical SCR fired DC power supply for production loads dropped from 100 to 1000 kilowatts to supplies ranging from 10 kilowatts to approximately 50 kilowatts.

Operators discovered that the use of auxiliary heat allowed them to heat the load to drive off water, oil, and other debris, and as a result, sputter cleaning became a minor problem. Heating the load to temperature could be done as rapidly as the material and part geometry itself would allow without distortion or cracking. The simplified, smaller conventional SCR fired DC power supplies became very cost effective, while the ion nitirding system itself with auxiliary heat (or, more properly, a conventional vacuum furnace equipped with a multipurpose capability of ion nitriding) became the preferred method of operation.

Little operator skill was required to operate the equipment beyond operation of a typical vacuum furnace. Skill was only necessary to establish treatment cycles for the load and the material in order to establish the nitride result, and to be able to fixture the load and place the thermocouple. Operator expertise was improved and, as a result, the confidence of the industry increased. Sufficient people worldwide were using the ion nitride process properly that acceptance of the process began to grow and become accepted by commercial heat treat shops, as well as by captive users.

As the controversy began to subside over the use of auxiliary heat versus non-auxiliary heat, pulsed high frequency power supplies began to be offered by equipment manufacturers with a new set of claims that pulsed high frequency power supplies offered significant improvements in the process of ion nitriding, as well as for in the ion nitriding equipment power supply.

The articles published on pulsed power supplies at the last conference reported that:

"A power supply for such a plasma nitriding plant has to fulfill the following conditions:

- Fulfillment of the physical conditions for abnormal glow discharge;

- Heating and temperature control of the workload for the heating-up period, and during treatment;

- Temperature uniformity in the workload and ;

- prevention of arc formation.

"With a conventional DC power supply, the fulfillment of these conditions is limited because the first three conditions were coupled. For the heating temperature control only a small gap of the abnormal glow discharge between the above-mentioned limits is available. Particularly at high temperatures this gap would be very small. Especially for a mixed work load with different geometries, the temperature uniformity is difficult to handle because small parts with a surface-to-volume ratio can be overheated. The minimum power input necessary for the abnormal discharge has to be balanced, e.g., by cooling the chamber walls, having a bad temperature uniformity as a result.

"The arc formation has to be detected and interrupted as fast as possible. The development from a glow discharge to an arc needs a certain time in the order of milliseconds. It is possible to cut off the current within this order, but even within this short time a small arc can damage the surface"... "A better solution to fulfill the above conditions is the use of the pulsed DC with some special conditions:

1) The form of the pulse should be a square so it is possible to jump from zero directly into the allowed gap of the abnormal glow discharge.

2) The pulsed length should be shorter than the development time of an arc, e.g., less than 100 microseconds, so that the arc formation is disturbed. If necessary, the interruption of the current should be possible during each pulse.

3) The following pause after pulse should be short enough to allow an easy ignition for the next pulse (e.g., less than some microseconds).

4) The ratio of the pulse to pause should be variable in a wide range to control the power input by the plasma in the workload so that it would be possible to use an auxiliary heating or better temperature distribution in the chamber."[1]

Grun concludes that with such a power supply a hot wall chamber design is desirable for providing temperature uniformity to the pieces so that the inner and outer pieces have "hot neighbors".[1]

This author does not feel that this is a significant factor for establishment of a pulsed high frequency power supply, as proper fixturing will always have inner and outer pieces seeing "hot neighbors" and such techniques to use fixturing and proper view factors have been used for over 30 years. Additionally, the use of conventional SCR fired DC power supplies in hot wall vessels has also been used for over 30 years and, as a consequence, the major factor in whether to use a hot wall or cold wall design revolves around proper equipment design and customer preference. It is not a necessary feature for pulsed or conventional SCR fired DC power supplies. Grun also claims that the pulsed technique reduces the strain caused by sputtering and the treatment of small bores is better, and control of the nitriding process is easier to manage. The reasons given for this are:

"The input of plasma energy into the workload can be dosed very fine within a wide range. Thus, a very good energy balance is achieved in the system with very low energy losses. It also results in a good temperature uniformity within the furnace on work pieces with different geometric shapes."[1]

A second author in the previous Ion Nitriding Conference (Kwon, et al) reported on a comparative study between pulsed and DC ion nitriding behavior in specimens with blind holes.[2] Kwon (et al) reported that the pulsed high frequency power supply allowed deeper penetration in holes over conventional DC ion nitrided holes.[2] They also reported that the growth rate of pulsed ion nitriding was slower on the outer surface of the test pieces, and faster on the inner surface of the test pieces. This article on face value would have provided definitive answers to questions raised in this paper except for the fact that careful review of the article indicated that two different sets of equipment were used in the tests, and two different methods of measuring temperature were also used in obtaining the test data.

The conventional DC power supply equipment utilized a chamber, 1000 mm in diameter x 1500 mm high, with temperature measured by an optical pyrometer. The pulsed high frequency tests were conducted in a chamber approximately half the size (400 mm x 600 mm high), and temperature measured by a thermocouple imbedded in a sample. Additionally, the gas mixture used was 66% nitrogen, 33% hydrogen. Under those circumstances, we would expect a greater preponderence of epsilon compound layer which would build up as a function of time and temperature, and depth of the diffused case would be governed by Ficks law of diffusion. Careful review of this paper could make an equally viable argument that the equipment and temperature could be the source of the result reported.

In this study, we attempted to eliminate the variables of the equipment and temperature measurement by utilizing the same equipment for both conventional SCR fired power supplies, as well as use of pulsed high frequency power supplies. As a manufacturer which offered both types of power supplies, as well as equipment with hot wall or cold wall, we had no particular prior result in mind which we were attempting to prove.

Initial studies were done with prototype pulsed high frequency power supply at Exactatherm, Ltd. (a commercial heat treating and ion nitriding service) in their facilities in Scarborough, Canada, and Missasauga, Canada. In that series of tests, we simply disconnected the conventional SCR power supplies from the equipment on site, and connected the pulsed high frequency power supplies. The original test in Scarborough was on a DC only heated laboratory system. The larger production unit in Missasauga utilized auxiliary graphite heating elements. Approximately two years later, Exactatherm ran a second series of tests on two bell style ion nitriding systems which were originally supplied with conventional SCR fired DC power supplies and subsequently converted to the use of pulsed high frequency power supplies. These units utilized only DC power

for heating. Exactatherm ran another series of tests which subsequently converted the small laboratory test system from conventional SCR fired power supply to a pulsed high frequency power supply. Most recently, Elatec ran another series of tests in conjunction with Nitron, another commercial ion nitriding service located in Lawrence, Massachusetts. In this facility Elatec had supplied a conventional vacuum annealing furnace equipped with an SCR fired DC power supply. (The auxiliary heaters in this system were molybdenum rod heating elements.) For these tests we had the opportunity to directly switch to pulsed high frequency power supply on production loads. As a result of our observations of all of these series of tests, we are prepared to report the following:

1) In circumstances with equipment using auxiliary AC heat, under normal load conditions with parts with normal geometry (a recognizably vague statement), there is no noticeable advantage to using either type of power supply. In effect, either system will work quite well and no benefits or advantages to either system could be really reported.

2) The circumstances when extremely dirty parts were treated without prior pre-cleaning, a definite advantage could be claimed for the pulsed high frequency power supply. In the case without the use of auxiliary AC heat, the pulsed high frequency power supply was able to bring the load to temperature, while the conventional SCR fired power supply was unable to bring the load significantly above ambient temperature. Further, under these circumstances, considerable operator experience was required to attempt to continuously increase power without creating arc damage with the conventional SCR DC only heated power supply.

3) In circumstances using DC heat alone, in which hollow cathode discharge was encountered, the pulsed high frequency power supply was far easier for the operator to utilize and find an operating pressure regime in which he could extinguish the hollow cathode effect. The conventional SCR fired DC power supply required substantial operator interface to prevent temperature from falling, or the hollow cathode discharge effect from overheating or even melting the parts.

4) With the use of auxiliary AC heat, the majority of the problems with hollow cathode discharge and conventional SCR power supplies disappear as auxiliary heat is used to maintain load temperature

while a different pressure regime is sought
in which to treat the parts without the
hollow cathode discharge effect. A slight
advantage goes to the pulsed high frequency
power supply under these circumstances. In
loads which have extreme geometry such as
gears with different pitches, it was found
that the pulsed high frequency power supply
had several major advantages. In circumstances where the conventional SCR fired
power supply simply would not perform
adequately and cover the surface of the
gear load, the pulsed high frequency power
supply would cover the surface, allow the
operator to avoid hollow cathode discharge
and provide a uniform metallurgical result.
The pulsed power supply was able to normalize
the energy distribution across the load
allowing the proper metallurgical result to
occur.

5) In tests run on sample pieces with blind
holes, it was found that the duty cycle
adjustment present in a pulsed high frequency
power supply could prevent hollow cathode
discharge formation (at a constant temperature within a constant pressure). It was
also noted that once a stable hollow cathode
discharge had occurred in a blind hole, that
the duty cycle (on the equipment provided in
this test), had no effect on eliminating
the hollow cathode discharge.

6) In the test pieces we had provided, we noted
a positive influence of the pulsed high
frequency power supply in that the frequency
of pulses had a great effect on arc suppression, but little effect on a penetration on
blind holes. (It is possible that a pulsed
power supply with higher frequency of pulses
would show a significant effect, but this
was not investigated.)

In summary, we can state that the selection of
the power supply for the general range of application of the ion nitriding process is a matter
of personal preference. Systems equipped with
auxiliary heating minimize the operator skill
required to run the process. The selection of
a pulsed high frequency power supply and
auxiliary heating provides a greater range in
processing variables which can have a positive
benefit under "extreme" treatment conditions.

References

1. Grun, Reinar, "Pulsed Plasma Treatment, The
   Ionnovation for Ion Nitriding", Ion Nitriding,
   ASM International, pp 143-148.

2. Kwon, Lee and Yoo, "A Comparative Study
   Between Pulsed and D.C. Ion Nitriding
   Behavior in Specimens with Blind Holes",
   Ion Nitriding, ASM International, pp 77-82.

Advantages Of Auxiliary Heating With Ion Processing

- Minimizes the size of D.C. power supply required

- Minimizes or eliminates problems of sputter cleaning

- Minimizes problems of operating out of "Hollow Cathode Discharge" without reloading the furnace

- Reduces operating skills required the system

- Minimizes "view factor" problems in fixturing

FIGURE 1

Arcs

- Cause surface damage to parts if not properly controlled

- SCR fired D.C. power supplies are voltage controlled and if an arc is not detected and suppressed peak surge currents at several times the power supplies full load rating can occur

- Pulsed high frequency D.C. power supplies are current controlled, if an arc or short circuit occurs there is no surge in output current

FIGURE 2

Arc Detection Circuits For SCR Fired D.C. Power Supplies

- Maximum current limit (I max)

- Current increase (di/dt)

- Voltage drops (dv/dt)

FIGURE 3

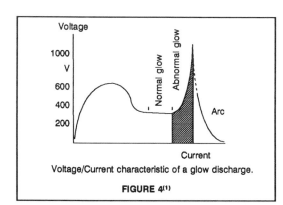

Voltage/Current characteristic of a glow discharge.

FIGURE 4[1]

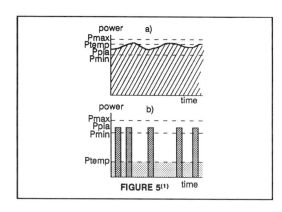

FIGURE 5[1]

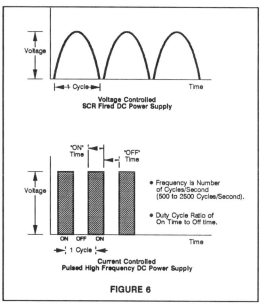

Voltage Controlled
SCR Fired DC Power Supply

- Frequency Is Number of Cycles/Second (500 to 2500 Cycles/Second).
- Duty Cycle Ratio of On Time to Off time.

Current Controlled
Pulsed High Frequency DC Power Supply

FIGURE 6

Pulsed High Frequency DC Power Suplly Specification

- Output: 1000 volts open-circuit at nominal input voltage Pulsed DC models available with maximum peak current ratings from 50 amps to 500 amps

- Control: 0 to 5 milliamps, 1250 ohms maximum 4 to 20 milliamps, 100 ohms maximum 10 to 50 milliamps, 100 ohms maximum

- Output Frequency: 500 to 2500 pulses per second

- Output Duty Cycle: 5% to 95%

- Maximum Line Variation: -20% to +10% of rated nominal input voltage

- Current Controlled

FIGURE 7

Special features of Pulsed High Frequency DC Power Suplly Tested

- Glow Interrupt Switch
- Current Control Potientiometer
- Automatic/Manual Current Control Switch
- Output Duty Cycle Potentiometer
- Pulse Frequency Potentiometer
- Arc Threshold Potentiometer
- Arc Suppression Time Potentiometer

FIGURE 8

Constants

- Temperature
- Pressure
- Load
- Furnace

Variables

- Type of power supply
  - Pulsed high frequency Dc
  - Conventional SCR fired DC

FIGURE 9

Hollow Cathode Discharge

Duty cycle adjustment can prevent HCD formation

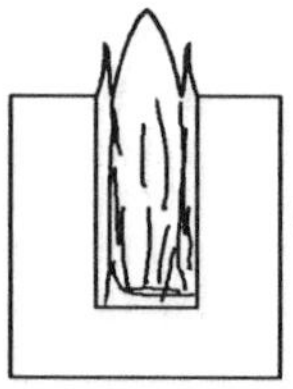

Once stable in HCD

Duty cycle had no influence

Constant temperature
Constant pressure

FIGURE 10

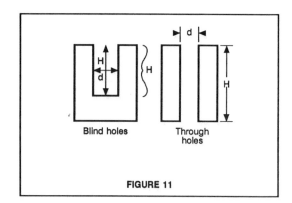

Blind holes

Through holes

FIGURE 11

Frequency of pulses had a great effect on arc suppression.

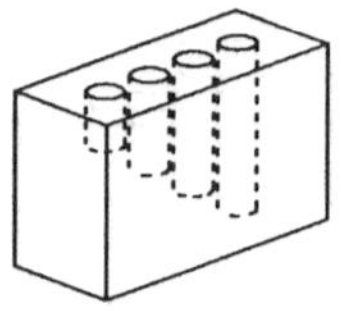

Frequency of pulses had little effect on penetration of hole.

Diameter of hole constant depth of hole varied temperature constant pressure constant.

FIGURE 12

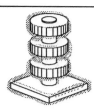

Large surface area constant DC power builds charge causing stable arcs to form with arc damage as a result.

Collapsing field with pulsed DC reduces arcing by allowing even energy distribution over load. Higher voltage and power without arc damage.

FIGURE 13

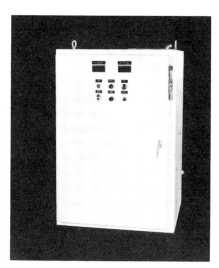

FIGURE 14
SCR Fired D.C. power supply used in tests at Nitron.

FIGURE 16
Belljar system used for second series of tests at
Exactatherm, LTD, Missasauga, Ontario, Canada
(manufactured by Elatec, Inc.).

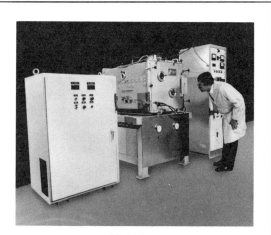

FIGURE 15
System used for tests
(courtesy of Nitron, LTD, Lawrence, MA).

FIGURE 17
Pulsed high frequency power supply used for tests
(shown with microprocessor control system used in
second series of tests at Exactatherm, Missasauga,
Ontario, Canada) (manufactured by Elatec, Inc.).

FIGURE 18
Multipurpose Plasvac System built by Elatec rated at
1650°C (second hot zone capable of using oxidizing
gases rated to 1200°C) (manufactured by Elatec, Inc.).

FIGURE 20
Small laboratory system used in first and second series of tests
at Exactatherm, Scarborough, Ontario, Canada (manufactured
by Surface Combustion, retrofitted by Elatec, Inc.).

FIGURE 19
Small Belljar system used in second series of tests at Exac-
tatherm, Missauga, Ontario, Canada, (manufactured by Surface
Technology, retrofitted by Elatec, Inc.).

FIGURE 21
Test load under glow with convention SCR fired D.C. on
Nitron system (showing thermocouple(s) and auxiliary
heating system).

COST EFFICIENT HIGH VOLUME PRODUCTION

Erich Hochbichler
Ion-Tech Inc.
300 Edgeley Blvd.
Concord, Ontario
Canada L4K 3Y3

Introduction

The application of Ion Nitriding can contribute to substantial cost savings for the manufacturing industry.

Avoiding distortion through processing and the improvement of wear properties vs. traditional heat treated components will provide the following cost saving benefits:

Production.Components

-Shorter manufacturing cycles and faster work flow;
-Completing all machining (incl. grinding) prior to processing reduces overall machining time (less machining and only soft materials to be removed);
-Components are ready for assembly as received from processing, secondary operations (post cleaning, dimensional checks, sorting, straightening or rust inhibitation) can be eliminated;
-Improved overall performance will result in increased competitiveness;
-Less scrap due to dimensional deficiencies.

Tools

-Extended tool life cycle will reduce tooling costs and breakdowns or maintenance cycles;
-Less material pick up and damage risk.

The cost savings on the above depend on:

-complexity of components and distortion sensitiveness;
-wear application;
-processing costs for ion nitriding;

The following subject will focus on Processing Costs for Ion Nitriding, which are in dependance of the process equipment and its features.

COST EFFECTIVE HIGH VOLUME PRODUCTION

The major considerations that need to be taken in order to decide upon utilizing the ion nitriding process for production components are improved wear properties and especially for production components, the achievable manufacturing cost savings. Most production components are already produced by using other heat treating methods and require specification change procedures. Therefore, it is essential to be able to offer cost savings or competitiveness to provide the necessary incentives to involved industries to implement the superior ion nitriding process. It should be avoided that gained manufacturing cost savings are lost for increased material and processing costs.

Ion nitriding process equipment is capable of producing costs that are competitive with other heat treating methods provided the equipment allows its optimum utilization and loading capacity. Since the capital costs to operate ion nitriding systems are about 40% to 50% of the operational budget, optimum utilization is the key to economical processing.

Optimum utilization means that any time other than actual process time (ion nitriding) is to be reduced to a minimum. This can be accomplished primarily by selecting equipment features to enable such efficient utilization.

The following variation of ion nitriding equipment concepts will show the impact on operational costs and available capacity.

COST AND CAPACITY STUDY

This cost and capacity study is based on an actual average process time (ion nitriding at temperature) of 8 hrs. and considers increased material (electricity and gas) and labour requirements in accordance to capacity variations. Overhead costs (administration, lease, and indirect labour etc. will remain the same for all equipment options listed.

-The cost of operations are indicated in: Cost/hr. and per furnace load/batch;
-The capacity is indicated by batches (furnace loads) per day.

EQUIPMENT FEATURES:

A1:   -1 Vacuum vessel: Dia. 48" x H 60" (15.7 cf.),
      Mat. 304 Stainless Steel;
      -1 Vacuum system (pump and controls);
      -1 Pulse power generator: 100 KW;
      -1 Process control system (computerized)

BUDGET PRICE $ 370,000.00

B1:   -2 Vacuum vessels: Dia. 48" x H 60" (31.4 cf.),
      Mat. 304 stainless steel;
      -2 Vacuum systems (2 pumps and 2 controls);
      -1 Pulse power generator: 100 KW;
      -1 Process control system (computerized)

BUDGET PRICE $ 465,000.00

C1:   -1 Vacuum vessel: Dia. 72" x H 80" (31.4 cf.),
      Mat. 304 stainless steel;
      -1 Vacuum system (2 pumps and 1 control);
      -1 Pulse power generator: 200 KW;
      -1 Process control system (computerized)

BUDGET PRICE $ 545,000.00

A2:   -Vacuum vessel: Dia. 48" x H 60" (15.7 cf.),
      Mat. 304 Stainless Steel;
      -1 Vacuum system (pump and controls);
      -1 Pulse power generator: 75 KW;
      -1 Process control system (computerized);
      -1 Auxiliary convection heating system: 75 KW
      and gas circulation turbine.

BUDGET PRICE $ 415,000.00

B2:   -2 Vacuum vessels: Dia. 48" x H60" (31.4 cf.),
      Mat. 304 Stainless Steel;
      -2 Vacuum systems (2 pumps and 2 controls);
      -1 Pulse power generator: 75 KW;
      -1 Process control system (computerized);
      -2 Auxiliary convection heating systems: 2 x 75 KW
      and gas circulation turbines.

BUDGET PRICE $ 510,000.00

C2:   -1 Vacuum vessel: Dia. 72" x H 80" (31.4 cf.),
      Mat. 304 Stainless Steel;
      -1 Vacuum system (2 pumps and 1 control);
      -1 Pulse power generator: 150 KW;
      -1 Process control system (computerized);
      -1 Auxiliary convection heating system: 120 KW and
      gas circulation turbine.

BUDGET PRICE $ 565,000.00

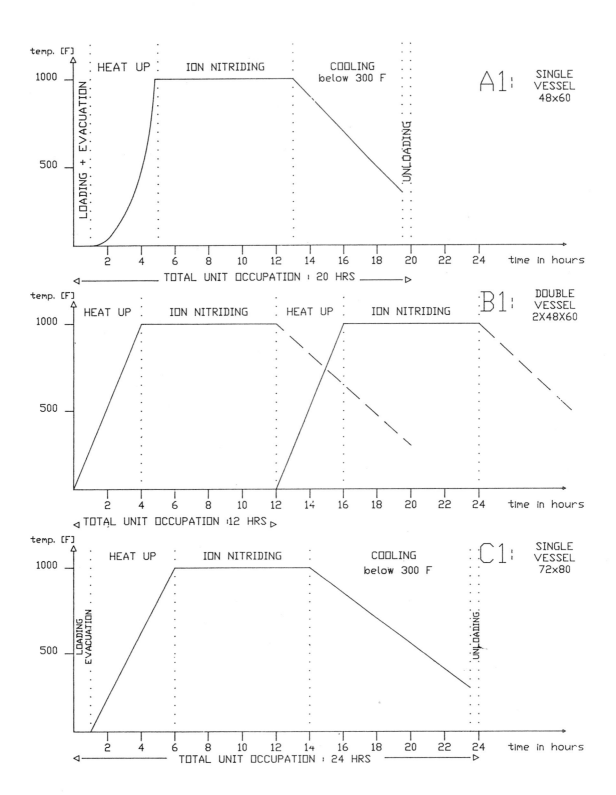

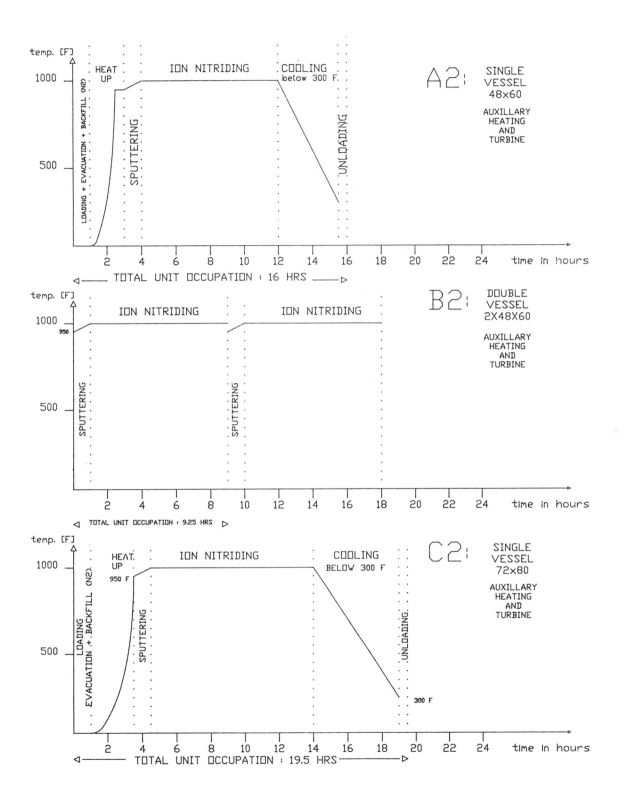

A1:	$ 370,000.00		Annual	Per Hour (6,000 hrs/ yr. or 250 days)
	Interests:	12.5%	$ 46,250.00	$ 7.71
	Depreciation:	5 years	$ 74,000.00	$ 12.33
	Labour		$ 42,000.00	$ 7.00
	Materials		$ 36,000.00	$ 6.00
	Overhead		$ 75,000.00	$ 12.50
			$ 273,250.00/yr.	$ 45.54/hr.

(20 hrs.) COST PER BATCH/LOAD: $ 910.80

B1:	$ 465,000.00			
	Interests:	12.5%	$ 58,125.00	$ 9.69
	Depreciation:	5 years	$ 93,000.00	$ 15.50
	Labour		$ 60,000.00	$ 10.00
	Materials		$ 60,000.00	$ 10.00
	Overhead		$ 75,000.00	$ 12.50
			$ 346,125.00/yr.	$ 57.69

(12 hrs.) COST PER BATCH/LOAD: $ 692.30 (x)

C1:	$ 545,000.00			
	Interests:	12.5%	$ 68,125.00	$ 11.35
	Depreciation:	5 years	$ 109,000.00	$ 18.20
	Labour		$ 48,000.00	$ 8.00
	Materials		$ 54,000.00	$ 9.00
	Overhead		$ 75,000.00	$ 12.50
			$ 354,125.00	$ 59.02

(24 hrs.) COST PER BATCH/LOAD: $ 1,416.48

<u>A2:</u> $ 415,000.00

Interests:	12.5%	$ 51,875.00	$ 8.65
Depreciation:	5 years	$ 83,000.00	$ 13.83
Labour		$ 42,000.00	$ 7.00
Materials		$ 42,000.00	$ 7.00
Overhead		$ 75,000.00	$ 12.50
		$ 293,875.00	$ 48.98

(16 hrs.) COST PER BATCH/LOAD $ 783.68

<u>B2:</u> $ 510,000.00

Interests:	12.5%	$ 63,750.00	$ 10.63
Depreciation:	5 years	$ 102,000.00	$ 17.00
Labour		$ 72,000.00	$ 12.00
Materials		$ 72,000.00	$ 12.00
Overhead		$ 75,000.00	$ 12.50
		$ 387,750.00	$ 64.13

(9.25 hrs.) COST PER BATCH/LOAD $ 593.20

<u>C2:</u> $ 565,000.00

Interests:	12.5%	$ 70,625.00	$ 11.77
Depreciation:	5 years	$ 113,000.00	$ 18.83
Labour		$ 60,000.00	$ 10.00
Materials		$ 60,000.00	$ 10.00
Overhead		$ 75,000.00	$ 12.50
		$ 378,625.00	$ 63.10

(19.5 hrs.) COST PER BATCH/LOAD $ 1,230.45 (x)

(x)=Double load volume

SUMMARY OF COST AND CAPACITY STUDY

	A1	A2	B1	B2	C1	C2
Equipt. Type	1 vessel Dia. 48/60 100 KW	1 vessel Dia. 48/60 Aux. htg. + turbine 75KW	Dbl.vsl. 2 x Dia. 48/60 100KW	1 vsl. 2 x Dia. 48/60 Aux. htg. + turbine 75KW	1 vsl. Dia. 72/80 200KW	1 vsl. Dia. 72/80 Aux. htg. + turbine 150KW

Capital Investment

	A1	A2	B1	B2	C1	C2
	$370,000	$415,000	$465,000	$510,000	$545,000	$565,000

Capacity per day

	A1	A2	B1	B2	C1	C2
	1.2 lds.	1.5 lds.	2 lds.	2.5 lds.	1 ld.	1.23 lds. (dbl./qty.)

A dual process utilization (8 hrs.)/day.

	A1	A2	B1	B2	C1	C2
	40%	50%	66.7%	84.2%	33.3%	41.7%

Net Cost P. Hour (6,000 hrs/year)

	A1	A2	B1	B2	C1	C2
	$ 45.54	$ 48.98	$ 57.69	$ 64.13	$ 59.02	$ 63.10

Net Cost P. Load (8 hrs. nitriding time)

	A1	A2	B1	B2	C1	C2
	$ 910.80	$ 783.68	$ 692.30	$ 593.20	$1,416.48 (Double batch size)	$1,230.45

Profit Margin P. Day - Based on Same Lot Price of Parts Quoted for A-1
(33% = $1,211.00)

33%	68%	125%	216%	(118%)	(198%)

Increased Profit Potential

	+35%	+92%	+183%	(+85%)	(+165%)

The above calculations are based on the assumptions of maximum loading.

CONCLUSION:

Despite the higher capital costs, double vessel installations are much more cost
efficient to operate. Even single vessel units with double capacity are almost
competitive. The double vessel units provide more flexibility in meeting
production requirements. Product groups with different process requirements
(process parameters) receive an improved response time. Any maintenance work
(cleaning of vessel, oil change on vacuum pump, thermo couple or O-ring
replacement etc.) can be accomplished without production loss by double vessel
installations. The smaller vessel dimensions of double vessel installations vs.
one large vessel with the same capacity will provide easier access for loading
parts (or fixtures). The same accounts for temperature uniformity in dependance
of the type of product to be processed.

Commercial heat treaters ignoring the benefits of ideal equipment selection will
have a hard time staying competitive. Flexibility and cost efficiency are
essential to provide an ultimate service to customers at reasonable costs.

The more cost competitive the ion nitriding process that is being offered is, the
faster this new technology will find new customers.

The process with its superior benefits in regards to superior wear resistancy and
cost savings for the manufacturing industry, has the potential to become the
ultimate heat treatment choice.

SUMMARY

Ideal equipment selection is the key for cost effective production.

>   -COST EFFICIENCY
>   -RELIABILITY
>   -FLEXIBILITY

If the equipment meets the above requirements, ion nitriding can be accomplished
a very attractive costs.

The benefits: A superior product at less costs than obtained by
                 applying traditional heat treating processes.

ADVANCED HARDNESS MEASUREMENT TECHNIQUE FOR LARGE SURFACE TREATED COMPONENTS

David M. Jankowski
Krautkramer Branson
Lewistown, Pennsylvania, USA

THE EFFECTIVENESS OF AN ION NITRIDING OR CARBURIZING TREATMENT is often determined by the relative magnitude and shape of the generated hardness profile. A typical ion treated profile is shown in Fig. 1. The three critical regions of the profile which control the performance of the treated component are the near-surface hardness, the core hardness, and the case depth.

Ion treatments are most commonly performed to increase NEAR-SURFACE HARDNESS. A higher hardness leads to increased wear resistance and fatigue life. Sufficient CORE HARDNESS or strength is necessary to support stresses transmitted through the surface hardened layer. The CASE DEPTH is defined as the depth to a critical hardness level; for an ion treated component, the case depth is frequently the depth to a hardness level 2 Rockwell "C" points above the core hardness.

Of these three parameters, the surface hardness is the most easily measured and is often used as a reliable indicator of the effectiveness of the ion treatment.

SURFACE HARDNESS MEASUREMENT

Several factors must be considered during surface hardness measurement. The indentation depth must be substantially less than the thickness of the hardened layer. A material thickness-to-indentation ratio of 7:1 to 10:1 is often recommended to minimize the effect of the core material on the measured hardness value. For a typical ion nitrided component with a surface layer depth of 0.025 to 0.102 mm (0.001 to 0.004 inch), the maximum indentation depth should be 0.005 to 0.015 mm (0.0002 to 0.0006 inch) deep.

Many ion treated components have finish machined or ground surfaces. Consequently, the depth of the hardness indentation should be minimized to eliminate the need and cost of corrective machining.

Large ion treated components such as automotive panel stamping dies may be difficult or impossible to move to a stationary hardness tester. Sectioning of the die is obviously not desirable. And sample coupons or shims included in the furnace batch may have significantly different properties than the die.

These three factors indicate the need for a low load, portable hardness tester.

MICROHARDNESS AND VICKERS HARDNESS TESTERS

Conventional microhardness and Vickers hardness testers apply loads ranging from 1 gram to over 5 kilograms. The diagonal (d) and depth (p) of the indentation can be easily calculated from Eqs. (1) and (2):

$$d = \sqrt{\frac{C_1 \bullet F}{H}} \qquad \text{Eq. (1)}$$

d = diagonal of indentation (mm)
C_1 = constant dependent on type of indenter
= 1.8545 for a Vickers diamond indenter
F = applied load (kg)
H = Vickers hardness

$$p = \frac{d}{C_2} \qquad \text{Eq. (2)}$$

p = depth of indentation (mm)
C_2 = constant dependent on type of indenter
= 7.5 for a Vickers diamond indenter

The major limitation of a conventional Vickers microhardness tester in portable on-site hardness measurement applications is the need to optically measure the indentation. A practical alternative is to use ultrasonic contact impedance measurement technology to evaluate the indentation.

ULTRASONIC CONTACT IMPEDANCE OPERATING PRINCIPLE

The ultrasonic contact impedance (UCI) operating principle was initially described by Kleesattel. It is based upon the relationship of the mechanical impedance between two elastic bodies and their contact area. In principle, a rod-shaped resonator is excited to its natural frequency by a pair of piezoelectric crystals. The rod is oscillated sinusoidally at approximately 80,000 Hertz. The fixed rod length limits the oscillations to the first harmonic effectively eliminating destructive interference. The resonant frequency of the rod increases when it contacts another object. The frequency shift is measured by a second pair of crystals mounted on the rod. The contact area can be determined from the measured frequency shift and the elastic constants of the rod and the test object. Eq. (3) defines the relationship.

$$A = g(E_d \nu_d E_p \nu_p) \bullet h (\Delta f/f_0) \qquad \text{Eq. (3)}$$

A = contact area
E_d = Young's modulus of rod
ν_d = Poisson's ratio of rod
E_p = Young's modulus of test piece
ν_p = Poisson's ratio of test piece
Δf = frequency shift
f_0 = initial frequency

The quantity $g(E_d \nu_d E_p \nu_p)$ can be determined by calibration with an appropriate standard or by optically measuring the contact area.

The mounting of a Vickers diamond indenter on the tip of the rod-shaped resonator and the application of a load to press the indenter into the test piece allows Vickers hardness measurement as defined by Eq. (4).

$$H = F/A \qquad \text{Eq. (4)}$$

H = Vickers hardness
F = applied load (kg)
A = contact area

After calibration for the elastic constants- the function $g(n)$ in Eq. (3) - the frequency shift defines the contact area and the hardness.

UCI BASED HARDNESS TESTER

A schematic of a UCI based hardness tester is shown in Fig. 2. The rod-shaped resonator is excited by a feedback amplifier through a pair of crystals. The Vickers diamond is mounted on the end of the rod-shaped resonator. A static load of approximately 800 grams is produced by constant load tension springs. The rod assembly is lowered and raised by an electric motor. A frequency discriminator measures the frequency shift. The load application time is adjustable from a minimum of two seconds to a maximum exceeding thirty seconds.

An operational UCI based hardness tester (Fig. 3) consists of the probe and the instrument. The probe has been previously described. The instrument contains the power supply, operating controls, electronic processing components, and display. The hardness can be displayed directly in Vickers or in converted Rockwell values.

EXPERIMENTAL RESULTS

A comparison of Vickers hardness values measured by a UCI based hardness tester and a conventional microhardness tester was conducted on a series of seventeen steels ranging in composition from a plain carbon 1010 to alloyed Cr-Ni-Mo 4340. The steels were in the quench and tempered or annealed heat treatment conditions. The UCI based hardness tester was calibrated with a 560 HVN block prior to the tests. The experimental results (Fig. 4) show an excellent 1:1 correlation between values measured with the two techniques. Kising et al. have shown similar results in the 550 to 800 HVN range.

HARDNESS MEASUREMENT OF ION TREATED COMPONENTS

A UCI based hardness tester allows surface hardness measurement of shallow cased ion treated components (Fig. 5). A surface layer hardened to 780 HVN (approximately 63.5 Rockwell "C") will have an indentation diagonal length of 0.041 mm (0.0016 inch) and an indentation depth of 0.0055 mm (0.0002 inch) when tested with a Vickers indenter and a 0.8 kg test load. The indentation depth is substantially less than the thickness of the hardened layer. UCI measurement eliminates the need for optical evaluation of the indentation. The small Vickers tip allows testing in radiused or difficult-to-access areas.

CONCLUSION

A UCI based hardness tester combines the benefits of ultrasonic contact impedance measurement technology and microhardness testing allowing hardness measurement of ion treated components. Specifically:
1) The relationship between contact impedance and indentation contact area has been established.
2) The contact impedance has been effectively measured with an ultrasonically excited rod-shaped resonator.

3) A UCI based hardness tester has been
manufactured by mounting a Vickers diamond on
a rod-shaped resonator and applying a constant
spring force to drive the diamond into the test
piece.

4) The shallow indentation of the UCI based
hardness tester and the portability of the test
unit allows on-site testing of large, difficult-
to-move objects such as ion nitrided stamping
dies.

REFERENCES

1. Kleesattel, C. and G. M. L. Gladwell,
    Ultrasonics, Vol. 6, 175-180 (1968).

2. Szilard, J., Ultrasonics, Vol. 10,
    174-178 (1984).

3. Kising, J., W. Weiler, and I. Winckler,
    "The UCI Principle", VDI Berichte, No. 583,
    371-391 (1986).

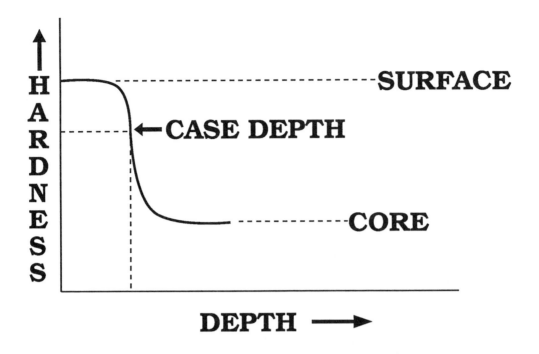

Fig. 1 - Hardness Profile of an Ion Treated
Component (Critical Regions Indicated)

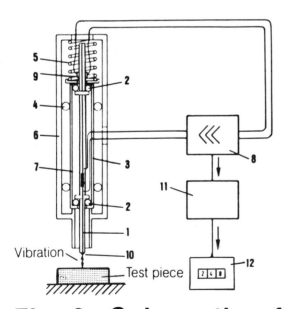

1. = oscillating rod
2. = rubber seal
3. = metal sleeve
4. = ball bearing bushings
5. = special spring
6. = housing
7. = oscillation converter
8. = amplifier
9. = oscillation detector
10. = diamond tip
11. = frequency discriminator
12. = ammeter

Fig. 2 - Schematic of a UCI Based
Hardness Tester

Fig. 3 - The MICRODUR - An Operational
UCI Based Hardness Tester

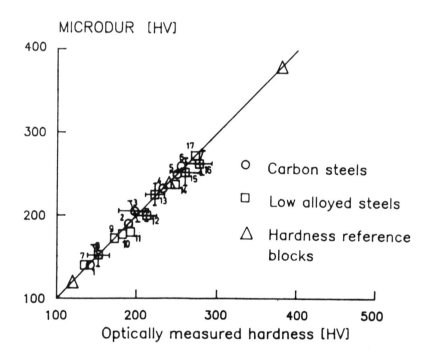

Fig. 4 - Comparison of UCI and Optically
Measured Hardness Results

Fig. 5 - UCI Based Hardness Testing of an Ion Nitrided Race

PLASMA (ION) CARBURZING, APPLICATIONS AND EXPERIENCES

Wolfgang Rembges, Jen Lühr
Klöckner Ionon GmbH
Leverkusen, FRG

Plasma(Ion)Carburizing,

Applications and Experiences

by Dr. W. Rembges, J. Lühr,
5090 Leverkusen 3, FRG

1. Introduction :

Environmental properties of heat treatment processes become more and more important. Quite a few of heat treatment facilities are working as if they are from the beginning of this century. They are using poisonous salts or produce harmful gases or smoke.

Nowadays there is the possibility to replace these processes with energy saving and clean technologies by introducing plasma heat treatment methods. Ionitriding[1] and Ioncarburizing of Klöckner Ionon belong to these technologies which reduce production costs and meet the modern requirements of reliability and flexibility.

The advantages of applying Ioncarburizing can be better appreciated by sharing some of the experiences of an aircraft component manufacturer and a contract heat tretment shop which both have installed a Klöckner Ioncarburizing production equipment.

---

1) Ionitriding[R], registered trade mark of Klöckner Ionon GmbH, FRG

2. Characteristics of the Plasma Heat Treatment Method: Ioncarburizing

The thermochemical surface hardening process Ioncarburizing (Plasmacarburizing) is performed in vacuum furnaces. The implementation of carbon takes place at a pressure range of up to 10 torr.

At a power density range from 0.1 to 1 Watt/cm² the transition coefficient of carbon is sufficiently high to enable a direct implementation of carbon into the surface of the workpiece. Thus, superior surface quality is achieved at a much shorter process time. Reduced energy consumption, fewer capital equipment together with high flexibility [6] and environmental impact are some of the distinguished features of this plasma technology.

The Ioncarburizing process as a case hardening process has been already known for several years but recently there has been an increased interest due to its remarkable process advantages and the availability of production type Ioncarburizing installations [1].

3. Ioncarburizing Installations

An Ioncarburizing furnace is similar to a two chamber vacuum-furnace. It consists of a load chamber with an integral quench chamber which is separated from the heat treatment chamber by a vacuum tight sealing door. The treatment chamber is equipped with the heating elements and the special device to create a glow discharge on the workpieces. Picture 1 shows the furnace in a heat treatment work shop, being used for daily production and different products.

Picture 1: Ioncarburizing unit installed in a work shop centre for heat treatment and process development

In this case the quenching tank is placed in a pit while the furnace can be loaded from the normal factory level. The control of the process is performed by a SMP[2]-Microprocessor system. The IONIT[3]-Software [2,4] controls and regulates the complete cycle. The comfortable software generates a simple and reproducible menu driven process. According to the customers experiences it is easy to change a program. The software is easy to handle by the operator. All important items are to see on the

CRT's. The system is equipped with an integrated trouble diagnosis system, which enables the operator to intervene quickly in case of a problem.

An Ioncarburizing cycle consists of the following steps: When the front chamber is loaded with the workpieces, it is pumped down to the same pressure as in the treatment chamber. With the help of a microprocessor controller, the load is transferred into the working chamber and is heated under vacuum to the required temperature. During heating up two soaking temperatures may be used to obtain uniform temperature throughout the load. After soaking at carburizing temperature the pressure of the working chamber is raised between 1 to 10 torr by adding controlled amounts of methane or other carbon containing gasmixtures.

The electrical power is connected to the workpieces (cathode), isolated from the chamber (anode) and the carburizing takes place under the control of the microprocessor. The carburizing cycle is followed by a diffusion step to reduce surface carbon. Surface carbon level and effective case depth can be achieved by interjection of several boost-diffuse cycles at the same temperature, too. During these cycles the microprocessor controls and regulates the sequence of the various process parameters.

At the end of the cycle the load is transferred to the quench chamber and is quenched in a vacuum-oil which is agitated by a fan to achieve the desired metallurgical properties. The load transfer mechanism is capable of providing a rapid transfer of the load from the hot zone to the quench tank.

As the process is performed under vacuum and the gases used are free of oxygen no internal grain boundary oxidation is observed in any case. Thus, tolerances normally left for cost intensive machining as in atmospheric processes are eliminated.

2) Siemens Microprocessor System (SMP)
3) IONIT[R], registered trade mark of Klöckner Ionon GmbH, FRG

4. Basic Process Control

The surface reactions in Ioncarburizing under glow discharge conditions are not well defined. Some results indicate that the reactions in the cathode fall near the surface are similar to those known in Ionitriding. Probably iron-carbide ions are created by sputtered iron atoms and then these ions are attracted to the workpiece surface area by the electrical field. Picture 2 describes schematically the reactions at the surface similar to those of Ionitriding [10].

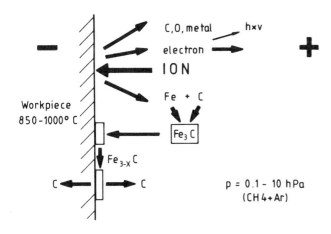

Picture 2: Reactions at the surface during Ioncarburizing

A significantly high carbon concentration (plasma C-potential) is created on the surface which is followed by carbon diffusion. With Ioncarburizing the plasma is used to ionize a reactive carbon plasma only. The heating up of the furnace is maintained by an auxiliary heating system, adapted to the requirements of the plasma heat treatment furnace size.

The described reactions in the plasma at the surface ensure an increased availability of carbon. This results in an overcarburizing of the surface layer while the plasma is on. The rate of carbon buildup at the surface is much higher compared to the conventional

carburizing. Thus, the carbon diffusion coefficient is increased which effects the carbon diffusion speed. In order to achieve a required surface carbon concentration the carburizing in the plasma has to be followed by a diffusion process.

This is easily attained as the temperature can be maintained without the glow discharge by means of the auxiliary heating elements. During the diffusion process the case depth is increased and the surface carbon concentration is depleted. The carbon diffusion takes place only into the metal and no carbon escapes into the vacuum atmosphere. Thus, an "S" profile of the carbon concentration is achieved.

First empirical results have shown that a ratio of 1 to 2-3 of the carburizing time to diffusion time gives the best surface hardness after direct quenching. A program has been developed to determine the ratio of carburizing to diffusion time in advance. This *IonCarburizing Calculation Program* (**Ionit-ICC-Programm**) calculates the exact treatment cycle corresponding to the required surface carbon concentration, core carbon content, material (DIN or AISI) and the treatment temperature on a personal computer [9].

After calculation of the process parameters, the Ionit-ICC program prints out the complete treatment recipe for the operator. It can be down loaded from the personal computer to the microprocessor of the Ioncarburizing unit via an interface line (RS 232).

The following example shows how the surface carbon concentration can be changed by varying the ratio of carburizing time to diffusion time:

Example:

Material: SAE 8620
Treatment temperature: 930°C ≡ 1705°F

Surf.-carb. : 0.7 wt%/ 0.8 wt%/0.9 wt%
Ioncarb-time: 35 Min/ 51 Min / 69 Min
Diff.-time : 148 Min/133 Min /115 Min

Picture 3 shows the carbon profile and the corresponding hardness profile of an ioncarburized sample after a treatment at a temperature of 900°C ≡ 1650°F. In this case the ratio of carburizing to diffusion time was 1:3.5. When the actual carbon concentration (0.67%) obtained at 0.05mm was compared to the value obtained from the Ionit-ICC program (0.70 %) this appears to be sound correlation. Data from several other applications have been used for verification and consequently Ionit-ICC program is presently used as a standard Ioncarburizing process determination [9].

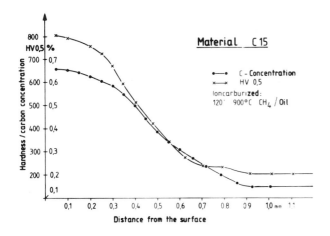

Picture 3: Carbon profile and hardness profile of an Ioncarburized sample made of Material C15/AISI 1015

Several already existing specifications on drawings are requiring surface carbon concentrations of more than .85% which may lead to an unaccecptable amount of retained austenite when direct quenching is used. This becomes especially important when higher temperatures are used with Ioncarburizing than with atmospheric processes in order to reduce cycle time. To influence the amount of retained austenite three different possibilities can be chosen:

a) lower quenching temperature after carburizing and diffusion cycle

b) deep freezing after direct quenching

c) change specification to lower surface carbon concentration

a) This methode is very simple but it has to be considered that the diffusion cycle has to be modified. By lowering the furnace temperature to the designated quenching temperature different times will be necessary according to the load sizes. During cooling down to the quenching temperature the carbon diffusion will continue and the surface carbon concentration is reduced while the case depth is increased. Thus the diffusion time has to be reduced appropiate in order to hold up the carbon concentration to the specified value.

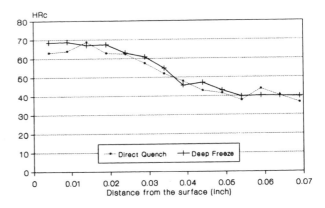

Picture 4: Hardness profiles of Ioncarburized samples, Material AISI 9310 after direct quench and deep freeze

b) Some materials have the tendency to form retained austenite in any case. In these cases it is recommendable to do deep freezing after quenching. Picture 4 and 5 show the hardness profiles of a test bars (AISI 9310) and a workpiece (AISI 4118) of an Ioncarburizing load after direct quench from 1700°F (926°C). The hardness drop near the surface is an indication of retained austenite. In both cases after an additional deep freeze in liquid Nitrogen the retained austenite is reduced to less than 3% and the hardness near the surface is increased.

This load was treated 120 min carburizing plus 140 min diffusion at 1700°F (926°C) and should have had a surface carbon of 0.92% and a case depth of 0.038 inch (0.97mm).

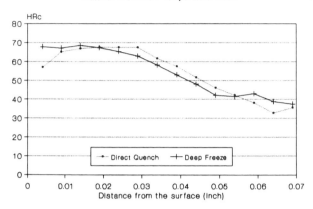

Hardness-Profiles
Direct Quench and Deep freeze
Material: AISI 4118 / Temperature 1700 F

Picture 5: Hardness profiles of Ioncarburized sample, Material AISI 4118 after direct quench and deep freeze

c) The above two possibilities ask for an additional operation which extend the cycle time and increases the production costs. Therefore it can be recommandable to change the specification and specify the surface hardness, the case depth according to 50HRc and the carburizing temperature instead of the surface carbon concentration and case depth according to 0.40% Carbon. Thus, the advantage of a reduced cycle time can be still used.

5. Materials for Ioncarburizing

Ioncarburizing is applied to all known cementation steels which can be quenched and hardened in oil. Additionally, the process can be applied to other steels based on the principle: activation of the carbon plasma of a glow discharge. As long as the materials are not highly alloyed the

carbon concentration profiles in the metal are not significantly influenced by the alloying elements but the hardness profiles are affected [7].

Examples of Ioncarburized steels show that the effective case depths of 550 HV ≡ 54 HRc follow a parabolic relation. Picture 6 shows along with typical carburizing steels results of a heattreatable steel 42CrMo4 ≡ AISI 4140 which was ioncarburized and quenched under the same conditions as the cementation steels.

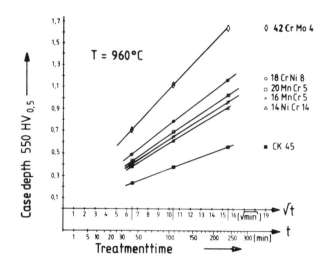

Picture 6: Parabolic growth of the effective case depth 550 HV0.5 ≡ 54 HRc for different cementation and one heat-treatable steel

Materials:

42 Cr Mo 4	≡	AISI 4140
18 Cr Ni 8	≡	AISI 3220
20 Mn Cr 5	≡	AISI 5120
16 Mn Cr 5	≡	AISI 5115
14 Ni Cr 14	≡	AISI 3316
Ck 45	≡	AISI 1045

Ioncarburizing can be extended to steels with higher content of Chromium. These group of steels have the tendency to form passive layers during conventional heat treat processes but in plasma technologies, the sputtering effect and the oxygen free carburizing atmosphere prevent formation of passive layers on the surface. Thus, Ioncarburizing overcomes the limitation based on the surface condition. The only limitation is the rate of diffusion within the layer. The advantages of this process are starting to be used for surface treatments of highly alloyed materials like M50NiL or stainless steels.

As the carbon implementation during Ioncarburizing is independent from the actual workpiece temperature, this technology can be applied over a wide range of temperature. As demonstrated in Picture 7 the same case depth can be achieved at different temperatures. By using higher temperatures, shorter cycle times are possible. To achieve a case depth of 1 mm ≡ 0.04 inch at 840 °C ≡ 1545 °F, 13.7 hours are necessary. At 930 °C ≡ 1705 °F the time is reduced to 4 hours, at 1020 °C ≡ 1870 °F only 1.4 hours are necessary. Increasing the carburizing temperature from 840 °C to 930 °C ≡ from 1545 °F to 1705 °F reduces the treatment time to 1/3. Thus, most of the carburizing cycles are performed at temperatures exceeding 930 °C ≡ 1545 °F. No significant grain size growth will be observed at 960°C ≡ 1760 °F if the cycle time is kept short.

process aircraft gears under its own personnel control. Picture 8 shows the unit which has been in operation trouble free for over one year. This furnace system consists of the same components as the before described one (see picture 1) but the chamber, the control and the power cabinets are designed according to A.S.M.E. standards.

The chambers are double wall, water cooled carbon steel construction with a full opening hinged door on both sides. The inner wall is coated with a rust preventative low outgassing coating to prevent oxidation and to facilitate cleaning. The furnace is supported on a platform, which is used for operating, loading and unloading of parts.

The useful size of the hot chamber is W: 24" * H:24" * L:36" and process loads up to 800 lbs. The hot zone construction is of modular design for easy disassembly and maintenance. The microprocessor software can operate in metric or in English system.

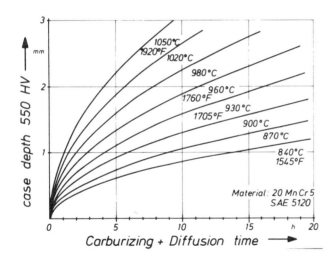

Picture 7: Growth of the ECD 550 HV 0.5 ≡ 54 HRC for different ioncarburizing temperatures, material 20MnCr5 / AISI 5120

Picture 8: Ioncarburizing unit, volume (H:24" * W:24" * L:36"), installed in an American automotive company

6. Practical Experiences

The advantages of the Ioncarburizing process mentioned above have encouraged an aircraft division of an American automotive company to purchase an Ioncarburizing unit from Klöckner to

The intrinsic heating elements and shielding are the major design requirements for a hot zone of a vacuum resistance furnace. A temperature uniformity test according to ASM 2750 was performed to meet the aerospace process requirements. The chart recorder in picture 9 shows the temperature uniformity of five thermocouples mounted to specimens, as

240

required. The temperature uniformity is better than ±10°F.

The quality of carburized parts is measured by the case depth, surface hardness, retained austenite and by case depth uniformity. In the case of gears the uniformity of the case along the tooth profile is of major concern. The different ratios of the surface area at the pitch and at the root to the interior volume influence the case depth uniformity at the pitch and at the root significantly in atmospheric processes. In a plasma process these differences can be minimized by creating a glow which follows the contour of the gear.

certain extend different pitch gears can be ioncarburized at the same time.

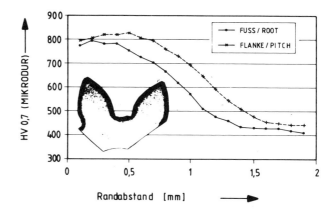

Picture 10: Hardness profiles and microsection of a gear with module 6.0 (4 D.P.), taken at the root and the pitch

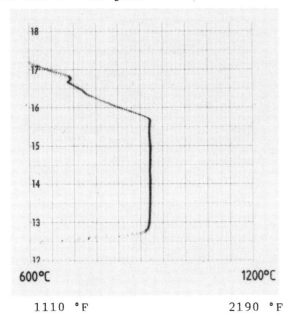

Picture 9: Temperature uniformity: recorder chart of five thermocouples mounted to control specimens in the Ioncarburizing hot zone

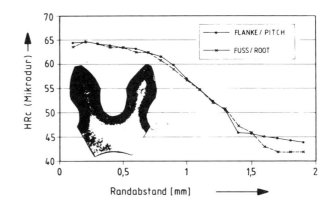

Picture 11: Hardness profiles and microsection of a gear with module 2.5 (10 D.P.), taken at the root and the pitch

The pictures 10 and 11 are typical examples which show the hardness depth curves and the corresponding microsection of gears of two different (2.5 and 6.0) modules (4 D.P. and 10 D.P.). Both type of gears were treated in the same load to attain a case depth of 1.2mm (0.045"). The fine pitch gear, 10 D.P., reveals no significant difference when measured against the 6 pitch gear. The ratio of case depth at the root to that of the pitch line is 86% which is better in comparison to the atmospheric processes for the same pitch gears. The other significant fact is that up to a

By optimizing the parameters this difference can be minimized. Because the bending strength is specified by minimum case depth at the root Ioncarburizing allows to shorten the cycle time. An additional advantage will be that no tendency to "Overcase" the pitch and tooth tips will occur and machining tolerances normally left for extra case removal can be avoided.

241

These results of aircraft gears correspond to the positive experiences with different types of gears which have been Ioncarburized in a work shop. As an example bigger gears each weighing 115 kg ≡ 255 lbs manufactured of material 15CrNi6 (AISI 3215) are carburized for 300 minutes to a case depth of ≥ 1.4 mm ≡ 0.055" at a temperature of 960 °C ≡ 1760 °F.

Internal holes or other areas can be protected against carburizing by a solid masking. After quenching and tempering the surface hardness inside the hole or at the protected will be according to the hardness of the not carburized core hardness. In case of the material 15CrNi6 (AISI 3215) it is about 40 HRc.

Picture 12: Load of piston in front of a plasma carburizing furnace, threads and internal holes protected by solid masking

Picture 12 shows a load of pistons made of 20MnCr5 (AISI 5120). The threaded areas are masked by a tube and the internal holes are automatically masked by the way of loading the parts on a flat metal sheet. This is another typical example of an application of Ioncarburizing in a job shop. By using this process several maching costs are omitted and no aftermaching operation is necessary because the carburized

martensitic layer is free of internal oxydation.

Besides the improved metallurgical results of ioncarburized parts the distortion can be important to the application of this process, too. Deformations of case hardened workpieces are mainly produced during the quenching operation. With plasma carburizing there is no influence to the quenching operation. Therefore it is more surprising that in several cases the distortion is less with Ioncarburizing. This can only be understand by the better uniformity of ioncarburized layers.

Distortion data measured with bearing cages are chosen as an example; the data are summarized in table 1. About 30 cages have been measured before and after a cycle in an atmospheric and in an Ioncarburizing production furnace. The average distortion and differences in roundness is significantly less with Ioncarburizing. The main advantage is the statistical deviation being 30% less for Ioncarburizing. Thus, the number of pieces which may be out of tolerance is reduced.

Table 1:

Treatment:	Roundness[*]		Stat.Dev.
	Before	After	±3σ
Gascarb.	0.076	0.0976	0.195
Ioncarb.	0.076	0.0822	0.136
Treatment:	Distortion[*]		Stat.Dev.
	Before	After	±3σ
Gascarb.	0.035	0.0692	0.155
Ioncarb.	0.035	0.0853	0.115

[*] All data in mm according to statistical measurements

7. Conclusion:

The experiences in using an Ioncarburizing unit in a heat treatment work shop and at a consumer's place in the aircraft industry show that in both cases this technology can be applied to different products and workpieces. The advantages of the process can be summarized as follows:

- reduced cycle times compared to atmospheric processes

- higher temperature cycles can be used without influencing the life of the furnace

- consistent and uniform case depths

- Ioncarburizing can be extended to other type of steels

- oxidation free surface layers, thus eliminating final machining

- reduced distortion in several cases

- uniform temperature distribution in the hot zone

- no sooting problems

- reduced energy and very low gas consumption

8. Literature:

[1] Grube, W.L. et al., High Rate Carburizing in a Glow Discharge Methane Plasma, Met. Trans. A, Vol 9 (1978), H.10 page 1421 - 1429

[2] Rembges, W., Oppel, W. Mikroprozessorsteuerung für die Wärmebehandlung, Gas Wär. Int. (1984) Bd. 33 Heft 6/7, S. 349 - 353

[3] Hombeck, F. Scientific and Economic Aspects of Plasmanitriding, Proceedings Volume, Shanghai 1983

[4] Rembges, W., Oppel, W. Mikroprozessorsteuerung für die Wärmebehandlung, Gas/Elektro-Wärme International 6/7, 84

[5] Hombeck, F. Forward View of Ion Nitriding Applications, Proceedings of an Intern. Conf. on Ion Nitriding Cleveland, Sept.1986, page 169

[6] Rembges, W. Fundamentals, Applications and Economical Considerations of Plasma Nitriding, Proc. of an Intern. Conf. on Ion Nitriding, Cleveland, Sept.1986, p. 189

[7] Wyss, U. Kohlenstoff- und Härteverlauf in der Einsatzhärtungsschicht verschieden legierter Einsatzstähle HTM 43 (1988) 1, Seite 27 - 35

[8] Rembges, W. et al. Berichtsband Einsatzhärten, AWT-Tagung 12.-14.Apr. 1989, Darmstadt, S. 93 - 106

[9] Rembges, W. Ionit-ICC Program, Klöckner Ionon, private communication

[10] Kölbel, J., Die Nitridschichtbildung bei der Glimmnitrierung, Forsch. Ber. d. Land. NRW, Nr1555, (1965)

Acknowledgement

The authors would like to thank Madhu Chatterjee from Allison Gas Turbine Division, Indianapolis for his help and support by preparing this paper.

Authors

Dr. rer. nat. W. Rembges, Head of Process Engineering and Process Developement Dept. of Klöckner Ionon GmbH, 5090 Leverkusen 3 / FRG Director of Engineering, Klöckner Ionon of America, Charlotte, N.C.

J. Lühr, Process Engineering and Process Developement Dept. of Klöckner Ionon GmbH, 5090 Leverkusen 3/ FRG

PLASMA AND GAS CARBURIZING
OF FINE PITCH AISI 9310 GEARS

W. L. Wentland, J. Y. Yung
Sundstrand Advanced Technology Group
Rockford, Illinois, USA

ABSTRACT

A comparative study was performed between plasma and gas carburized AISI 9310 gears. Very fine pitch (32 D.P.) test gears were machined and each were carburized by both plasma and gas carburizing methods to an effective case depth of 0.008-0.015". The resulting uniformity of the case was compared in the root and pitch diameter locations. The plasma carburizing process showed a greater ability to drive a deeper case in the root areas providing a better case depth uniformity. The required case hardness was easily achieved by plasma carburizing and the resulting microstructure was fine, uniform, and exhibited no retained austenite. Single tooth gear testing was performed on both plasma and gas carburized gears in both the ground and unground conditions and comparisons are shown. As predicted, the plasma carburized gears have no intergranular oxidation, while the gas carburized gears in the unground condition had some intergranular oxidation. The single tooth gear testing also shows the effect of this intergranular oxidation on the fatigue performance of the test gears.

CARBURIZED GEARS are widely used in many aerospace applications, such as auxiliary gear boxes, integrated drive generators, and actuation systems. Carburized gears have an edge over nitrided, carbo-nitrided, and through-hardened gears by having high allowables in all of the following major material design considerations: bending fatigue strength, contact fatigue resistance, and fracture toughness. While the above design features of carburizing are desirable, however, there are several problems associated with the conventional gas carburizing process. They include long cycle times, intergranular oxidation, and the inherent distortion which occurs when processing at temperatures above 1650°F. More importantly, there is a lack in gas uniformity in areas of restricted gas flow, such as in the roots of fine-pitch gears which reduces case depth. These problems lead to increased manufacturing costs, primarily from post-carburizing grinding.

Plasma carburizing has shown major advances in overcoming some of the difficulties associated with conventional gas carburizing. Plasma carburizing uses a carbon-bearing gas, usually methane or propane, under a partial pressure to provide an ionized carbon plasma at the work surface. The plasma tightly adheres to the part and greatly accelerates the transfer of carbon from the gas phase to the steel surface. This normally allows better uniformity of carburizing for fine-pitch gears over the atmospheric process [1]. Benefits claimed by the plasma process include superior case structure and uniform case depths, reduced distortion, short cycle times with low energy consumption, and lack of intergranular oxidation inherent with a vacuum process.

This paper compares the effects of both plasma and gas carburizing. The key elements for comparison were the case microstructure (particularly its uniformity), and the bending fatigue properties as obtained by single tooth gear testing.

PROCEDURE

GEAR MANUFACTURE — The material chosen for this study was vacuum melted AISI 9310. This is the alloy of choice for most of our gear needs. The gear chosen was a spur gear with a 32-pitch gear form of configuration as shown in Figure 1. This fine-pitch gear was picked due to its difficulty in obtaining a uniform case in the root areas. These gears were hobbed and copper-plated (maskant) on the sides and tips prior to carburizing. The profile requirements after grinding were an effective (50 HRC) case depth of 0.008-0.012" with a case hardness of 58-62 HRC at the pitch location and 54-62 HRC in the roots.

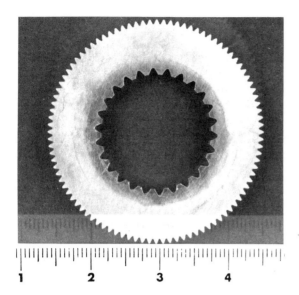

Figure 1. 32-Pitch test gear configuration

HEAT TREATMENT — The gas carburizing gears were carburized at 1550°F for 90 minutes in an endothermic atmosphere, and were subsequently cooled in a protective atmosphere until they could be properly handled, after which they were air cooled to room temperature. The plasma carburizing gears were carburized at 1562°F for 60 minutes (two alternating cycles of 10 minutes boost followed by 20 minutes diffuse) in a partial pressure of 3 torr with an argon/propane backfill and gas quenched to room temperature. The final heat treatment for all gears was as follows:

1. Sub-critical anneal at 1225°F for 2 hours, cool to room temperature.
2. Austenitize at 1515°F for 45 minutes, minimum.
3. Oil quench at 130°F.
4. Deep freeze at –110°F for 2 hours, and warm to room temperature.
5. Temper at 350°F for 2 hours.

Important note: the plasma carburized gears were austenitized in the plasma carburizing furnace in an inert atmosphere such that intergranular oxidation was avoided.

INSPECTION AND POST-PROCESSING — Gears of each type were inspected metallurgically and hardness profiles were run in the root and pitch locations. Half of the gears used for single tooth gear testing were left as-received, while the other half were ground. All of the gears to be tested were tempered appropriately to obtain a root hardness of 54 HRC so that the hardness would not affect the results.

FATIGUE TESTING — The single tooth bending fatigue test is used to generate S-N curve documentation using a limited number of gear specimens. This method has proven to produce consistent and reproducible results[2]. The fatigue tests were performed on a single tooth gear test rig where the gear is keyed onto a strain gaged shaft coupled to a reaction arm on which the load is applied via a hydraulic cylinder. The load reacted via the test tooth on a reaction block. The height of the reaction block is adjustable to insure the maximum bending stress. The fatigue tests were conducted at room temperature in load control (R-ratio = 0.1) at 5-40 Hz.

RESULTS

MICROSTRUCTURE — The microstructure of the plasma carburized gear is shown in Figure 2. Its case exhibited no intergranular oxidation, some fine and scattered carbides (to 0.0006" depth), and a tempered martensite structure with no visible retained austenite. The microstructure of the gas carburized gear is shown in Figure 3. By contrast, its case had more, but slightly finer carbides (to 0.0015" depth), a similar tempered martensitic structure, with intergranular oxidation present to a depth of 0.00015".

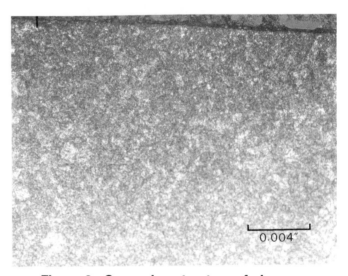

Figure 2. Case microstructure of plasma carburized gear

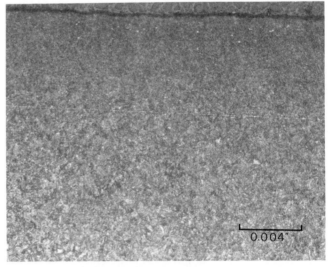

Figure 3. Case microstructure of gas carburized gear

CASE HARDNESS AND UNIFORMITY — The results of the carburizing trials can be found in Table 1. The first plasma carburized cycle, which was direct quenched, had excellent uniformity and hardness (Figure 4), but a non-ideal microstructure. The final plasma sample (Figure 5) had lower hardness and uniformity than the gas carburized sample (Figure 6). It appears that the time spent at the austenitizing temperature excessively diffused in the carbon thereby lowering the maximum hardness and uniformity while increasing the case depth. Decreasing the diffusion time should allow the results to approach that of the direct quenched sample.

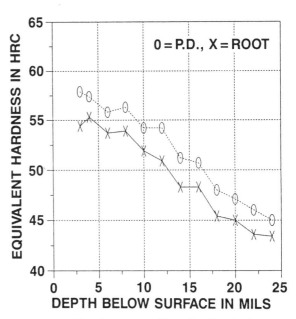

Figure 5. Hardness profiles for final plasma carburized gear

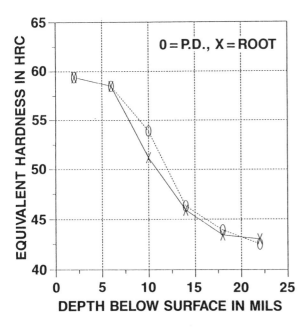

Figure 4. Hardness profiles for direct quenched plasma carburized gear

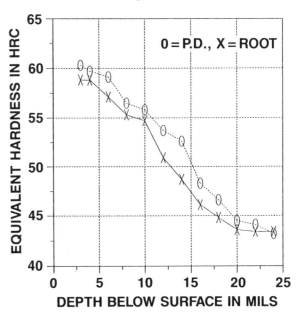

Figure 6. Hardness profiles for gas carburized gear

S/N	Carb. Type	Carb. Time (Min.)[1]	Max. Hardness		Case Depth (in.)		Root/P.D. Ratio
			P.D.	Root	P.D.	Root	
356	Plasma	10C + 20D + 10C + 20D	59.4	59.4	.012	.0108	89%
G-3	Plasma	10C + 20D + 10C + 20D + A	57.9	55.3	.0166	.0127	77%
T-3	Gas	90C + A	60.3	58.8	.0152	.0128	84%

Notes: 1. C = Carburize, D = Diffuse, A = Reaustenitize
2. All Hardnesses in HRC

Table 1. Carburizing results on 9310 32-pitch gears

SINGLE TOOTH GEAR TESTING — The results of the fatigue tests can be found in Figure 7. There was no significant difference between the plasma and gas carburized gears in the ground condition. Both of the unground gears, however, had significantly lower properties. Some of the drop is real, while some of the drop can be attributed to the less than ideal root profile produced by hobbing. A slight notch in the root radius definitely affected the results in that the fatigue cracks all initiated at this notch. Without the notch the plasma carburized unground gear should approach the same results as the ground plasma gear. The lower results in the unground gas carburized gear, when compared to the unground plasma carburized gear, shows the effect of the intergranular oxidation.

CONCLUSIONS

1. Plasma carburizing met the microstructural and hardness requirements on a 32-pitch AISI 9310 test gear. The microstructure of this gear was similar to a gas carburized gear except that no intergranular oxidation was produced as it was with the gas process.
2. Plasma carburizing was able to produce a more uniform case on this fine pitch gear. Although only a direct quench sample produced the more uniform case, a reaustenitized sample should significantly improve in uniformity with less diffusion time during carburizing.
3. There was no significant difference in the bending fatigue properties of both plasma and gas carburizing in the ground condition.
4. Plasma carburizing offers a higher bending fatigue allowable over gas carburizing in the unground condition. This provides a benefit if lower cost unground gears can be used in some applications. Proper comparisons could not be made with the ground gears due to difference in the root fillet radii.

REFERENCES

1. Conybear, J., Heat Treating, 24-27, March 1988.
2. Buenneke, R.W., "Gear Single Tooth Bending Fatigue Test", SAE Technical Paper No. 821042.

(a)

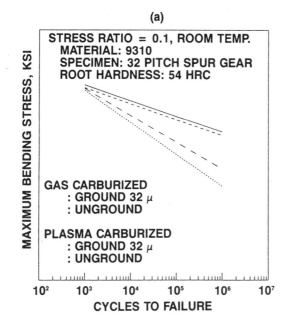

(b)

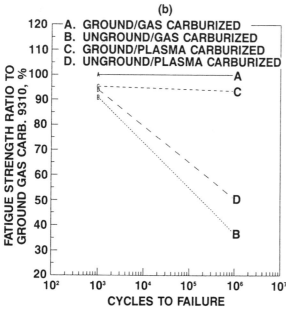

Figure 7. Single tooth gear fatigue (a) data and (b) comparisons for ground and unground plasma and gas carburized AISI 9310

PRACTICE AND EXPERIENCE WITH PLASMA CARBURIZING FURNACE

Koichi Akutsu, Masanori Nakamura
Daido Steel Co., Ltd.
Nagoya, Aichi, Japan

Abstract

Plasma carburizing, which is one of the cold plasma assisted surface modifications, is discussed now. Plasma carburizing is characterized by its oxidation free, soot free heat treatment and high carburizing ratio. This process is useful for high carbon carburizing, the case hardening of hard - carburizing materials such as stainless steels, high carbon alloyed steels and so forth.

To apply this plasma carburizing in practical use, Daido Steel Co., Ltd. has developed two types of furnaces. One is composed of two working vacuum chambers, one for plasma carburizing and one for gas or oil quenching. The other one is single chamber type with high pressure gas quenching facilities. Both furnaces are industrial size. The study, using these newly developed furnaces, has proved that plasma carburized Cr - Mn sintered parts show higher fatigue properties than conventionally carburized parts. Austenitic stainless steel, which is hard to be carburized by conventional gas furnace because of its surface oxide film, exhibits extreamly high case carbon content up to 3.0 % and high hardness of HmV 500. Furthermore, super carburized 0.15C - 5.0Cr - 0.19Mo steel shows finely precipitated carbides in the case with carbon content 2.0 %. Results show superior pitting durability to conventionally carburized steels.

RECENTLY MACHINERY PARTS have been compact, aiming at weight reduction and an improvement in reliability and strength is required by users.

The major problem is to improve fatigue ,wear and pitting resistance for automobile parts. Whenever parts such as bearings are processed, it takes too long to carburize.

Furthermore, in vacuum carburizing furnaces operated at pressures between 40 and 67 kPa, a heating chamber is covered with soot which is known to prevent the accurate carbon control. In conventional gas furnaces, where sooting has also occured, the carburizing time is 1.3 to 1.5 times as long as that in plasma.

In order to solve the above mentioned problems and satisfy users' needs, plasma carburizing can offer the many advantages as follows:-
   a. shorter processing time
   b. soot free
   c. no intergranular oxidation
   d. ability to carburize materials which are hard to be carburized by conventional gas furnaces
   e. automatic operation
   f. no environmental pollution

The purpose of this paper is to illustrate two types of newly developed plasma carburizing furnaces and to present practical results recieved in then. All of the experiments have been carried out with industrial - scale furnaces. Therefore, the results and experiences can be applied to a commercial operation.

FURNACE DESIGNS AND PROCESSES

A plasma carburizing furnace with two chambers is shown in Fig. 1 schematically. Fig. 2 shows an industrial-scale furnace which is installed Daido Steel Co., Ltd. This furnace has the working space dimensions of 900 mm long x 600 mm wide x 500 mm high, and can

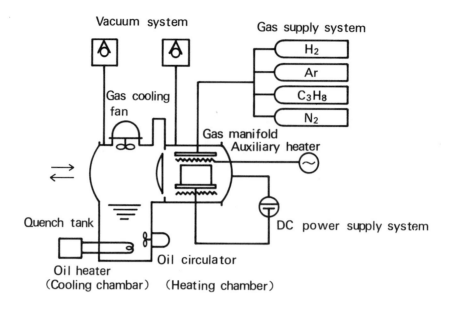

Vacuum system

Gas supply system

Gas cooling fan

H₂
Ar
C₃H₈
N₂

Gas manifold
Auxiliary heater

Quench tank

DC power supply system

Oil heater
(Cooling chambar)

Oil circulator
(Heating chamber)

Fig. 1 - Scheme of plasma carburizing furnace

load maximum gross charge weight of 450 kg. The cooling chamber is separated from the heating chamber by a vacuum-tight gate valve, and contains the transfer mechanism and the gas or oil quenching facility. This chamber provides loading and unloading operations. The plasma chamber is located inside a large heating chamber and consists of graphite auxiliary heaters, a heat insulator made of a carbon fiber, a manifold feeding carrier gases into a working space and a cathode table which is insulated from the plasma and heating chambers with earth potential.

In the plasma chamber, prior to carburizing, workpieces are automatically positioned on a cathode table. The plasma power levels available are 20 to 30 kVA for the industrial - scale units.

Another type of plasma carburizing furnace is equipped with high pressure gas quenching facilities as shown in Fig. 3. This furnace is characterized as achieving a homogeneous hardening, because the cooling gas gets everywhere. The working space dimensions in plasma chamber are 750 mm long x 500 mm wide x 400 mm high and maximum gross charge weight is 250 kg. The cooling gas

Fig. 2 - Industrial - scale plasma carburizing furnace with two chambers

Fig. 3 - Industrial - scale plasma carburizing furnace with single chamber

pressure available is maximum 698 kPa. To treat the materials to be oxdized easily and promptly, it is essential that a diffusion pumping system is prepared. And this furnace is fully automated with the aid of the computer which provides complete process control and monitoring of all process variables as shown in Fig. 4.

The processing sequence of two chambers type of plasma carburizing furnace involves the following several steps.

After loading up, the cooling chamber is evacuated to 7 Pa after the heating chamber has been previously evacuated, and the inner vacuum - tight gate valve is opened for transportation of workpieces into the heating chamber. Workpieces are heated to temperature in range of 780°C to 1050°C by auxiliary heaters without glow discharge. After heating up, soaking and sputtering in Ar and H$_2$ glow discharge are carried out. Case hardening steels don't need sputtering, prior to plasma carburizing, but it is essential to clean by sputtering in the case of stainless steels, which form the surface oxide film. In the next step, workpieces are plasma carburized. Propane plasma is essential if unformity of carburized case is required. After a programmed time, the glow discharge power is cut off, and diffusing period takes place up to the extent at which the required case depth and carbon content are achieved. However, the diffusing period is not necessary in the case of super carburizing. Carburized and diffused workpieces are transported in the cooling chamber and lowered into the oil tank.

Vacuum system, such as plasma carburizing furnace, is most suitable to operate automatically. The complete automatic production line as shown in Fig. 5 makes the most use of this advantage. This line consists of plasma carburizing furnace, tempering furnace, degreasing or washing equipment and loader/unloader device. For example, Degreasing - Plasma carburizing - Process annealing - Secondary quenching - Washing - Tempering processes can be carried out in one cycle.

In plasma carburizing, its specifications such as carbon content and case depth depend on treating temperature, carburizing time and diffusion time. Experimentally, it was found that case depth and carbon content were ruled by Harris' relationship as shown in Fig. 6 and 7.(1) In an industrial - scale plasma carburizing furnace, if treat-

ment pressure and DC current density are set up at a threshold value and above, it is not essential to control their parameters strictly, because surface carbon content achieves austenitic carbon solubility in a short time in propane plasma as shown in Fig. 8(2), and plasma power doesn't contribute to the temperature rising of workpieces.

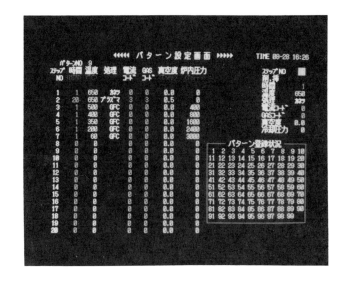

Fig. 4 - CRT display for plasma carburizing furnace

Fig. 5 - Automatic line for plasma carburizing

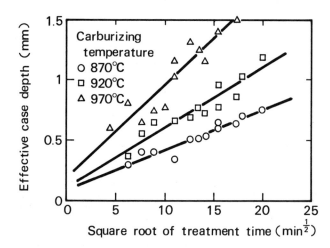

Fig. 6 - Harris' relationship at several temperatures of effective case depth as a function of square root of treatment time

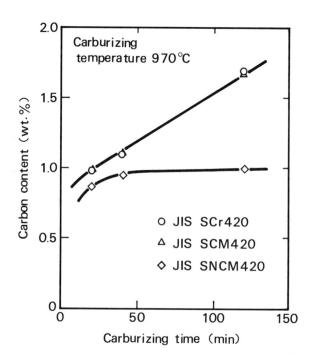

Fig. 8 - Changes in surface carbon content with carburizing time

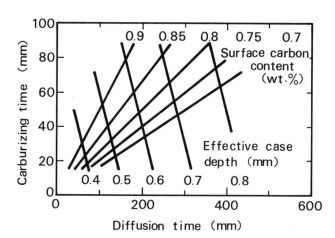

Fig. 7 - Relation between the carburizing and diffusion time determining on the desired carbon content and case depth

PLASMA CARBURIZING OF SINTERED PARTS

Sintered parts carburized by conventional gas furnaces are deeply carburized through core, because treatment pressure is atmospheric. This causes toughness to decrease. In sintered parts containing Cr and Mn, which are oxidized easily, the carburizing atmosphere allows the materials to oxidize not only at the surface but also into the core. On the other hand, in vacuum through hardening, carbon diffuses into sintered parts, and hardness gradient is not obtained. As a result, it is difficult to expect the comprehensive residual stress and to maintain the high toughness.

Plasma carburizing gives high strength to Cr - Mn sintered parts because of its lack of intergranular oxidation and glow discharge into fine pores. Therefore, a successful hardness profile is obtained as shown in Fig. 9. Fig. 10 demonstrates machanical properties of Cr - Mn sintered parts, which are applied to automobile parts, carburized at 950°C for 25 minutes by propane plasma and diffused for 100 minutes. Fig. 11 shows that the surface and inside of pores are not oxidized.

PLASMA CARBURIZING OF SUPER CARBURIZING STEEL

Super carburizing is a high - carbon carburizing process in which spheroidal carbides are finely precipitated in high carbon cases, such as 1.5 to 2.0 %

Fig. 12 shows carbon profiles of case hardening steels super carburized at 930°C within 16 hours' carburizing

time. Spheroidal carbides finely precipitate near the surface as shown in Fig. 13.

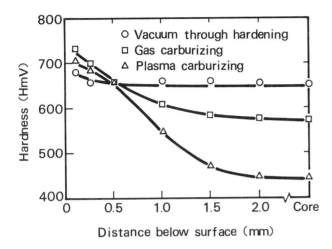

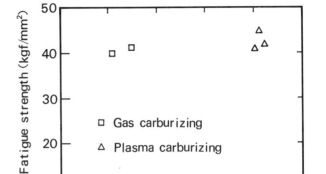

Fig. 9 - Hardness profiles of Cr - Mn sintered parts

Fig. 10 - Comparison of machanical strength of Cr - Mn sintered parts

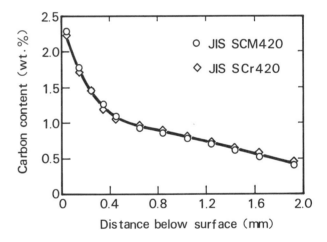

Fig. 11 - Microstructure of plasma carburized Cr - Mn sintered parts

Fig. 12 - Carbon profiles of JIS case hardening steels super carburized by propane plasma

Table 1 - Chemical compositions of JIS case hardening steels

	C	Si	Mn	P	S	Cu	Ni	Cr	Mo	sol.Al
JIS SCr420	0.21	0.22	0.78	0.014	0.010	0.08	0.05	1.13	0.02	0.034
JIS SCM420	0.20	0.26	0.82	0.021	0.017	0.09	0.08	1.16	0.15	0.031

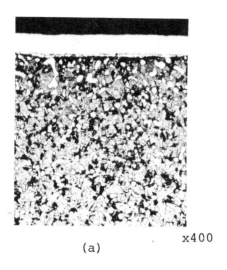

x400

(a)

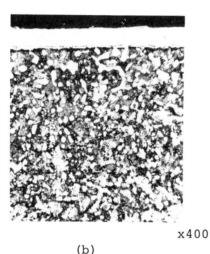

x400

(b)

Fig. 13 - Microstructures of
super carburized JIS case
hardening steels
(a) JIS SCM420
(b) JIS SCr420

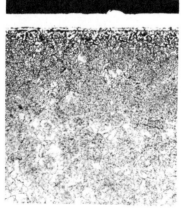

x400

Fig. 14 - Microstructure of
super carburized 5 % Cr
steel

Super carburizing steel containing larger amounts of carbide formers, especially Cr, is more suitable to high carbon carburizing. Fig. 14 shows the microstructure of super carburizing steel containing 5 % Cr after carburizing at 920°C for 3 hours. Its carbon and hardness profiles are also plotted in Fig. 15. In this case, retained austenite was 9 %. As shown in Fig. 16, it was found that plasma carburized specimen had much superior pitting resistance to conventional gas carburized ones.

PLASMA CARBURIZING OF STAINLESS STEELS

Stainless steels are hard to be carburized using conventional gas furnaces because of thier surface oxide film. In plasma carburizing, surface oxides can be reduced by sputtering with Ar and H_2 plasma prior to carburizing. Carburized stainless steels are used for parts which require corrosion resistance and nonmagnetism. However, carburized parts, as shown in Table 2, have far inferior corrosion resistance to nonprocessed ones. Therefore, regions of parts not requiring wear resistance were mechanically masked by solid materials, such as steel pipe, steel plate and carbon sheet.

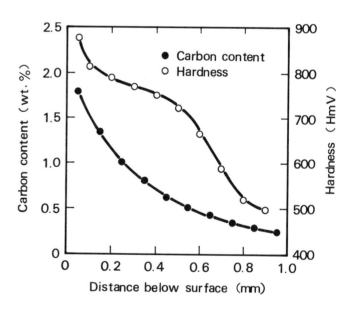

Fig. 15 - Carbon and hardness
profiles of super carburized
5 % Cr steel

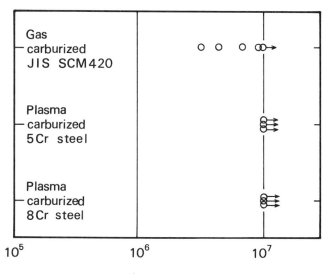

Contact stress : 375 kgf/mm²
Number of revolution : 1500 rpm
Slip ratio : −40%

Fig. 16 - Results of rotating -
bending fatigue test for super
carburized steels

Plasma carburized stainless steels were applied to a part of a bicycle, a precision machine and a food manufacturing machine. The hardness and carbon profiles of various stainless steels plasma carburized at 950°C for 7 hours are plotted in Fig. 17 and 18, respectively.

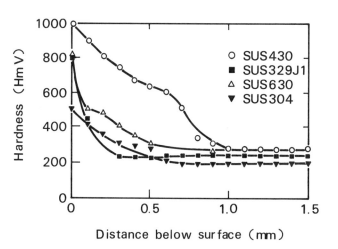

Fig. 17 - Hardness profiles
of plasma carburized
stainless steels

		Electrochemical potentiokinetic reactivation ratio (%)
JIS SUS 304	Nonprocessed	7.9
	Plasma carburized	88.9
JIS SUS 316	Nonprocessed	0
	Plasma carburized	56.3

Table 2 - Comparision of
corrosion resistance in
plasma carburized stainless
steels and nonprocessed ones
(Test method : JIS G 0580)

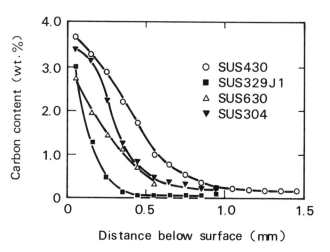

Fig. 18 - carbon profiles
of plasma carburized
stainless steels

255

ANOTHER EXAMPLE

Bearings require to be carburized to various depths depending on the application. Especially, large bearings carburized from 2 to 4 mm result in the improvement of the wear and spall resistance. At the same temperature, processing time of plasma carburizing is four - fifths as long as that of conventional gas carburizing. This proves that plasma carburizing is characterized by adhesion of C ions due to potential drop close to the cathode (the workpieces). As a concrete example, JIS SNCM 815 is plasma carburized at 980°C for total processing time of 38 hours to obtain surface carbon content of 0.9 % and effective case depth of 4.0 mm. Under the same temperature, it took 46 hours to obtain the same carburizing specifications in conventional gas carburizing.

CONCLUSIONS

This paper introduced several examples and experiences. The specific features of plasma carburizing are summarized as follows:-
   a. faster processing time
   b. upgrading of the product quality
   c. development of producing products with special feature
   d. automatic operation
   e. no pollution problems
As a result, plasma carburizing gives commercial heat treaters several new aspects to the treatment of case hardening steel, stainless steel, sintered parts and super carburizing steel.

It is concluded that plasma carburizing is expected to be the new surface modification method in the cases of hardening of nonferrous metals and complex materials.

ACKNOWLEDGEMENT

The authors are much indebted to Messrs. Y. Nagata, T. Matsuzawa and K. Tamamoto of NIPPON DENSI KOGYO Co., Ltd. for carrying out the experiments.

REFERENCES

(1) Harris, F. H., Met. Prog., 265-272 (1943)
(2) Kimura, T., K. Namiki, Current Advances in Materials and Processes, 1,(1988)

THE FATIGUE CHARACTERISTICS OF ION-SULFO-CARBONITRIDED STEELS

Xun Luan, Zhenglin Li, Buqian Wang
Mechanical Engineering Department
Xian Institute of Highways
Xian, P.R.C.

Abstract

Three low alloy steels: 20CrMnTi,30CrMnSiA and 35CrMo have been ion-sulfo-carbonitrided by the new technique,which we proposed.The fatigue characteristics of treated steels have been investigated and a comparison of the fatigue strength of ion-sulfo-carbonitrided steel with those of liquid sulfo-carbonitrided and gas-tuff-nitrided steels was made.

The results indicated that after ion-sulfo-carbonitriding,the depth and the strength of diffusion layer can be increased,the origin of fatigue crack was moved into the core and the fatigue strength,the overload endurance can be increased more effectively than other two processes. Moreover,ion-sulfo-carbonitrided specimens displayed a lower sensitivity to the effect of stress raisers on the fatigue strength.

A VARIETY OF SULFO-CARBONITRIDING PROCESSES have been widely used for improving both wear resistance and fatigue characteristics of ferrous materials.(1-5) However,in conventional process,liquid or gaseous agents are used,they have some disadvantages such as pollution and deterioration of the salt bath. We have thus developed a low temperature ion-sulfo-carbonitriding process using ion bombardment in the glow discharge vapor of mixed ammonia-alcohol-carbon disulfide. The new process has particular advantages over gaseous or salt bath treatments, namely shorter treatment time,energy economy,simple operation,good reproducibility and no environmental pollution(6).

It is the purpose of the present paper to investigate the fatigue characteristics of 20CrMn Ti,30CrMnSiA and 35CrMo steels treated by this new process.The microstructure of the surface and diffusion layer were observed and the microhardness profiles of treated samples,the fatigue crack initiation site as well as the strength of diffusion layer were analyzed and discussed. At the same time a comparison of the fatigue characteristics of ion-sulfo-carbonitrided steels with that of liquid sulfo-carbonitrided and gas-tuff-nitrided steels was made.

EXPERIMENTAL DETAILS

1) SAMPLE MATERIALS- The materials tested were three low alloy structural steels that are used in industry:20CrMnTi,30CrMnSiA and 35CrMo steels. Their compositions are listed in Table 1. The 20CrMnTi steel samples were in normalized state and 30CrMnSiA,35CrMo steel samples were in the hardened and tempered state prior to treatment.

2) ION SULFO-CARBONITRIDING PROCESS-The ion-sulfo-carbonitriding process was carried out in the glow discharge vapor of mixed ammonia-alcohol-carbon disulfide using the method we proposed.(6) The detailed processing parameters including the resulting layer thickness and the surface hardness of samples are shown in Table 2. For comparison ,the treating parameters and results of liquid sulfo-carbonitrided and gas tuff-nitrided samples all listed in Table 2.

3) STRUCTURAL PROPERTY EXAMINATION- The microstructure of surface and diffusion layers was examined by optical microscopy. The microhardness profiles of treated samples were measured using a microhardness tester with load of 0.98 N. The fracture surface of specimens and the origin of fatigue crack were observed by scanning electronic microscopy.

4) FATIGUE TESTING- The fatigue tests were carried out on a rotating-bend fatigue machine,at an uniform frequency of rotation of 3000 r.p.m.(7). 20CrMnTi steel specimens were machined into 6 mm in diameter and 80 mm in length and had a class 9 surface finish. For comparison test specimens were prepared in ion-sulfo-carbonitrided,liquid sulfo-carbonitrided and gas tuff-nitrided states. The smooth and notched 30CrMnSiA,35CrMo steel samples were prepared in both a tough hardened state and an ion sulfo-carbonitrided state and the smooth specimens were 6 mm in diameter and 80 mm in length. In order to determine the influence of stress raisers on the fatigue strengh of the ion-sulfo-carbonitrided steels,we cut grooves 1 mm deep and 4 mm wide in notched specimens 8 mm in diameter. The data obtained were for 10^7 cycles.

TABLE 1
Steel Composition(wt.%)

Steel	C	Si	Mn	Mo	Cr	Ti	P	S	Ni	Fe
20CrMnTi	0.17-0.24	0.20-0.40	0.80-1.10	--	1.00-1.30	0.06-0.12	<0.04	<0.04	--	Balance
30CrMnSiA	0.27-0.34	0.90-1.20	0.80-1.10	--	0.80-1.10	--	0.018	0.013	0.06	Balance
35CrMo	0.32-0.40	0.20-0.40	0.40-0.70	0.80-1.10	0.80-1.10	--	<0.04	<0.04	--	Balance

TABLE 2
Heat Treatment, layer thickness and hardness of specimens

Steel	treated state	process parameters			thickness		surface hardness(Hv100)
		T(°C)	t(hr)	$NH_3/C_2H_5OH+CS_2$	compound (μm)	diffusion (mm)	
20CrMnTi	I-SNC	560	3	20:1	20-30	0.40	700-770
	L-SNC	570	3	--	10-25	0.25-0.30	620-650
	G-T-N	560	5	--	7-15	0.15-0.25	460-480
30CrMnSiA	I-SNC	510-520	4	20:1	17.2	0.30-0.35	620
	T						285
35CrMo	I-SNC	550-560	4	20:1	19.5	0.25-0.30	493
	T						213

I-SNC: Ion sulfo-carbonitrided state, L-SNC: Liquid sulfo-carbonitrided state.
G-T-N: Gas tuff-nitrided state. T: Tough hardened(hardened and tempered)

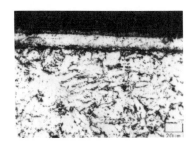

a) 20CrMnTi

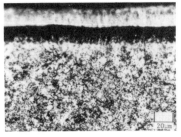

b) 35CrMo

Figure 1 The microstructure of the cross sections of ion sulfocarbonitrided 20CrMnTi(a) and 35CrMo(b) steels.

RUSULTS AND DISCUSSION

1) STRUCTURE OF THE LAYER- Figure 1 (a) (b) show the microstructure of the cross sections of ion-sulfo-carbonitrided 20CrMnTi and 35CrMo steels, respectively. It can be seen that the layer on the ion-sulfo-carbonitrided steel specimens consists of two parts: the outer compound layer-white layer(15-30 μm thickness) and the inner diffusion layer(0.3-0.4 mm thickness). There is some porosity in the outer part of compound layer, the next outer part of the compound layer is more compact and firmer. This corresponds to the microhardness profiles of the cross section as shown in Fig. 2. Outer surface exhibited a slight lowering in hardness and the inner diffusion layer had the higher hardness and gentle gradient of it. The microhardness profiles of 20CrMnTi steel samples treated by three processes and those of ion sulfo-carbonitrided 30CrMnSiA,35CrMo steel samples are shown in Figure 2(a) and (b) respectively.

From Figure 2-(a) it can be found that the gas tuff-nitrided 20CrMnTi steel sample has the highest hardness and two sulfo-carbonitrided samples exhibit a slight lowering hardness in outer

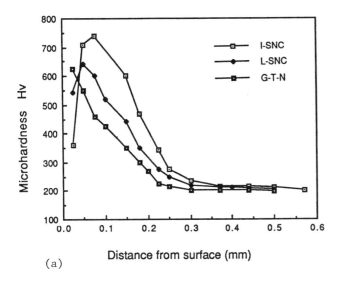

(a)

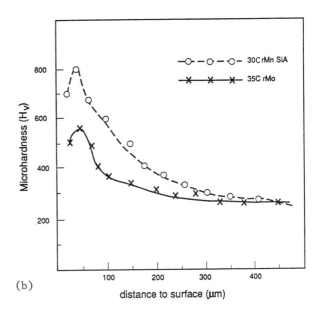

(b)

Fig.2(a)--The microhardness profiles of 20CrMnTi steel samples treated by three processes.
(b)--The microhardness profiles of ion sulfo-carbonitrided 30CrMnSiA,35CrMo steel samples .

surface ,but in subsurface ion sulfo-carbonitrided 20CrMnTi steel sample shows the highest hardness and a gentle gradient. In addition,Figure 2 (b) indicated that ion sulfo-carbonitrided 30CrMnSiA steel showed a higher hardness in layer than ion sulfo-carbonitrided 35CrMo steel sample.
2) FATIGUE STRENGTH- The S-N curves of 20CrMnTi steel specimens treated by ion sulfo-carbonitriding,liquid sulfo-carbonitriding and gas tuff-nitriding are shown in Figure 3.
From Figure 3 it can be seen that ion sulfo-carbonitrided sample showed the highest fatigue limit,the liquid sulfo-carbonitrided sample had the lower fatigue limit than the ion sulfo-carbonitrided sample and that of the gas tuff-nitrided sample was lower still. Combined Table 2,it is shown that the fatigue strength improvement is related to the case depth. Ion sulfo-carbonitri-

ded sample had the thickest depth,the fatigue limit was increased by 5.2% and 13% comparing with liquid sulfo-carbonitrided and gas tuff-nitrided samples,respectively. So with increasing case depth ,the fatigue limit increased,that is agreement with the finding of many investigators.(8-10)
Figures 4 and 5 show the S-N curves of 30CrMnSiA and 35CrMo steels,respectively,and indicated that the ion sulfo-carbonitriding process greatly increased the fatigue limit of both 30CrMnSiA and 35CrMo steels and for both notched and smooth specimens. The fatigue limit of 30CrMnSiA steel smooth specimens is increased by 50% and of notched specimens by 48%. Ion sulfo-carbonitriding increased by 34% the fatigue limit of 35CrMo steel for smooth specimens and by 43% for notched specimens. The increase in the fatigue strength of ion sulfo-carbonitrided steels is due to the high residual

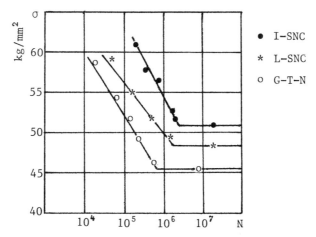

Figure 3 -- The S-N curves of 20CrMnTi steel specimens treated by ion sulfo-carbonitriding,liquid sulfo-carbonitriding and gas tuff-nitriding.

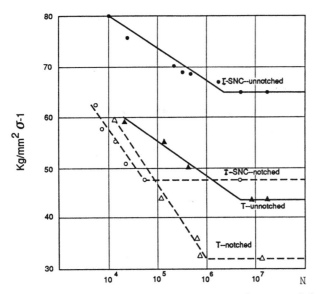

Fig.4 S-N curves of 30CrMnSiA steel treated by
I-SNC and by T(tough hardening.)

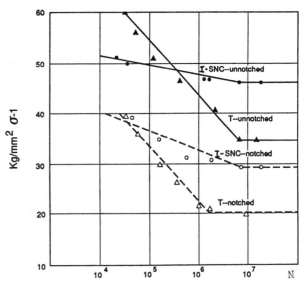

Fig.5 S-N curves of 35CrMo steel treated by
I-SNC and by T.

compressive stresses in the case and the thick depth and the high hardness of the diffusion layer.(1)(4)(8)

Table 3 shows the fatigue limit and K_f(the stress concentration factor of the notch on fatigue) of 30CrMnSiA and 35CrMo steel for both tough hardened and ion sulfocarbonitrided specimens. From Table 3 it can be seen that ion sulfo-carbonitriding reduced K_f for both steels tested and ion sulfo-carbonitrided 30CrMnSiA notched specimens exhibited a fatigue limit of 48 Kgf/mm² which is even higher than that (43 Kgf/mm²) of tough hardened 30CrMnSiA smooth specimens.Obviously the ion sulfo-carbonitrided steel specimens displayed a lower sensitivity to the effect of stress raisers on fatigue strength.
3) OVERLOAD ENDURANCE BEHAVIOR- From Figure 3 it can be seen that under the same stress(52.5Kgf-mm⁻²) ion sulfo-carbonitrided specimen had the life time of 1.2 X 10⁶ cycles, the life of the overload endurance increased by 440 % and 750 % comparing with liquid sulfocarbonitrided and gas tuff-nitrided specimens,respectively. From Fig.3 it also can be seen that the overload endurance curve of ion sulfocarbonitrided 20CrMnTi steel sample had the steepest slope than those of other two treated samples. Moreover, from Figs. 4 and 5 it can be shown that the overload endurance curve

of ion sulfocarbonitrided 35CrMo steel for both smooth and notched specimens had a slighter slope than those of untreated 35CrMo steel specimens, and the slopes of the curves of ion sulfocarbonitrided and untreated 30CrMnSiA steel for both notched and smooth specimens are very close. These indicated that ion sulfocarbonitrided 35CrMo steel had a higher sensitivity to overload than did the 30CrMnSiA steel. As shown in Table 2 and Figure 2-b, the ion sulfocarbonitrided 35CrMo steel had a thicker compound layer with a lower density and a steeper slope of the microhardness profile than did the 30CrMnSiA steel. Therefore,the fatigue microcracks that formed on the subsurface at an early stage are likely to propagate more easily for 35CrMo steel than for 30CrMnSiA steel, which appears to account for the lower overload resistance of ion sulfocarbonitrided 35CrMo steel.
4) FRACTOGRAPHY- all the facture surface of specimens were observed to have failed by the " fish eye " phenomenon with fatigue cracks originating from nonmetallic inclusions. As shown in Figure 6.

Combined Figure 6 with the microhardness profiles(see Figure 2-a),it can be clearly seen that the fracture initiation in the three treated specimens all occurred at the inclusion immediately below the case-core interface. Obviously,after ion sulfocarbonitriding the depth and the strength

TABLE 3

Fatigue limit and K_f

| Steel | State | Fatigue limit | | K_f |
		σ_{-1}(smooth) (Kgfmm⁻²)	σ_{-1n}(notched) (Kgfmm⁻²)	
30CrMnSiA	T	43.5	32	1.36
	I-SNC	65	48	1.35
35CrMo	T	34.3	20	1.7
	I-SNC	46.1	29	1.6

$K_f = \sigma_{-1}/\sigma_{-1n}$ is the stress concentration factor for a notch on fatigue.

a)I-SNC

b)L-SNC

c)G-T-N

Figure 6 -- The origin of fatigue crack of 20CrMnTi steel specimens
treated by three processes.(a-ISNC.b-L-SNC,c-G-T-N.)

of diffusion layer can be increased,comparison
with the other two processes.Therefore,the origin
of fatigue crack was moved into the core.As shown
in figure 6,in the ion sulfo-carbonitrided specimen,the crack initiation occurred at the place
which is below surface about 0.30 mm,while in the
liquid sulfocarbonitrided and gas tuff-nitrided
specimens those were below surface about 0.25 mm
and 0.15 mm,respectively. This corresponds to that
the ion sulfocarbonitrided specimen had the higher
fatigue strength than other two treated specimens.

Furthermore,the fracture surface of three
treated specimens is characterized by three zones:
an initiation site,a zone of crack propagation and
a region of final fracture.Which is a typical fatigue fracture surface. Figure 7 shows SEM fractographs of ion sulfocarbonitrided 20CrMnTi steel
specimen,illustrating the three distinct regions.
Figure 7-a shows the origin of the fracture,indi-
cating that fatigue crack was initiated at an inclusion,featuring a brittle fracture. Figure 7-b
shows the zone of fatigue crack propagation,illu-
strating that the core was failing by ductile fatigue and was characterized by chaotic striations
which was transgranular by a combination of ductile striation formation and void nucleation and
growth. Figure 7-c shows the region of final fast
fracture,illustrating that the final failure of
core by microvoid coalescence has occurred and
the dimples are present in a wide variety of size.
At the bottom of equiaxed dimples spheraidel par-

ticles of nonmetallic inclusions can be seen.

CONCLUSIONS

1) The ion sulfocarbonitrided layer consists of
two parts: the outer compound layer and the inner
diffusion layer.
2) The ion sulfocarbonitrided steel has a high
surface hardness with a gentle inward hardness
profiles. The depth and the strength of diffusion
layer can be increased,comparison with both liquid
sulfocarbonitriding and gas tuff-nitriding.
3) In the ion sulfocarbonitrided specimens the
origin of fatigue crack was moved into the inside
and the fatigue strength,the overload endurance
can be increased more effectively than other two
processes.
4) Ion sulfocarbonitrided steel specimens displayed a lower sensitivity to the effect of stress
raisers on the fatigue strength.

REFERENCES

1) G.V.Karpenk,V.I.Pokhmurovskii,V.B.Dalisov and
V.S.Zamikhovskii,"Influence of Diffusion on the
Strength of Steel," Trans.Tech,Aedermansdorff,1979
2) T.Bell,Metal Engineering,Quarterly 18(1976)1.
3) J.C.Gregory,Heat Treatment of Metals,2,55-64,
1972.
4) D.Michalon,G.Mazet and C.Burgio,Rev.Metall.,73
(10)(1976)711.
5) Buqian Wang,Xun luan,Zhenglin Li and Jun Zhang,
"Structure and Properties of Ion Sulfocarbonitrided Coatings on Several Structural Steels" Thin
Solid Films,166,281-290,(1988)

a)initiation site

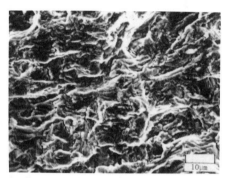

b)zone of propagation

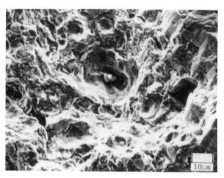

c)final fracture

Figure 7, SEM fractographs of ion sulfocarbonitrided 20CrMnTi steel specimen.

6) Buqian Wang,Ion Sulfocarbonitriding of Steel in the Vapor of Mixed Ammonia-Alcohol-Carbon di-sulfide,Proc.1st Int.Conf.on Plasma Surface Engineering,Garmisch-Partenkirchen,West Germany,Sept. 1988.
7) J.R.Newby,J.R.Davis,S.K.Refsner and D.A.Dieterich,Metals Handbook,Vol.8,Mechanical Testing, American Society for Metals.Metals.Park.OH.9th edn., 1985,pp.263,269
8) E.J.Mittemeijer,"Fatigue of Case Hardened Steels,Role of Residual Macro and Microstresses" J.Heat Treating,Vol.3,2,December,1983,114-119
9) B.K.Jones and J.W.Martin,"Metal Technology" Vol.5,7,217-221(1978)
10) T.Bell and N.L.Loh,"The Fatigue Characteristics of Plasma Nitrided Three PCT Cr-Mo Steel," J.Heat Treating,Vol.2,3,June 1982,232-237